The Advances publishes reviews and critical articles covering the entire field of normal anatomy (cytology, histology, cyto- and histochemistry, electron microscopy, macroscopy, experimental morphology and embryology and comparative anatomy). Papers dealing with anthropology and clinical morphology will also be accepted with the aim of encouraging co-operation between anatomy and related disciplines.

Papers, which may be in English, French or German, are normally commissioned, but original papers and communications may be submitted and will be considered so long as they deal with a subject comprehensively and meet the requirements of the "Advances".

For speed of publication and breadth of distribution, this journal appears in single issues which can be purchased separately; 6 issues constitute one volume.

It is a fundamental condition that submitted manuscripts have not been, and will not simultaneously be submitted or published elsewhere. With the acceptance of a manuscript for publication, the publisher acquires full and exclusive copyright for all languages and countries. 25 copies of each paper are supplied free of charge.

Die Ergebnisse dienen der Veröffentlichung zusammenfassender und kritischer Artikel aus dem Gesamtgebiet der normalen Anatomie (Cytologie, Histologie, Cyto- und Histochemie, Elektronenmikroskopie, Makroskopie, experimentelle Morphologie und Embryologie und vergleichende Anatomie). Aufgenommen werden ferner Arbeiten anthropologischen und morphologisch-klinischen Inhalts, mit dem Ziel, die Zusammenarbeit zwischen Anatomie und Nachbardisziplinen zu fördern.

Zur Veröffentlichung gelangen in erster Linie angeforderte Manuskripte, jedoch werden auch eingesandte Arbeiten und Originalmitteilungen berücksichtigt, sofern sie ein Gebiet umfassen abhandeln und den Anforderungen der „Ergebnisse" genügen. Die Veröffentlichungen erfolgen in englischer, deutscher oder französischer Sprache.

Die Arbeiten erscheinen im Interesse einer raschen Veröffentlichung und einer weiten Verbreitung als einzeln berechnete Hefte; je 6 Hefte bilden einen Band.

Grundsätzlich dürfen nur Arbeiten eingesandt werden, die nicht gleichzeitig an anderer Stelle zur Veröffentlichung eingereicht oder bereits veröffentlicht worden sind. Der Autor verpflichtet sich, seinen Beitrag auch nachträglich nicht an anderer Stelle zu publizieren. Die Mitarbeiter erhalten von ihren Arbeiten zusammen 25 Freiexemplare.

Les résultats publient des sommaires et des articles critiques concernant l'ensemble du domaine de l'anatomie normale (cytologie, histologie, cyto- et histochimie, microscopie électronique, macroscopie, morphologie expérimentale, embryologie et anatomie comparée). Seront publiés en outre les articles traitant de l'anthropologie et de la morphologie clinique, en vue d'encourager la collaboration entre l'anatomie et les disciplines voisines.

Seront publiés en priorité les articles expressément demandés, nous tiendrons toutefois compte des articles qui nous seront envoyés dans la mesure où ils traitent d'un sejet dans son ensemble et correspondent aux standards des « Revues ». Les publications seront faites en langues anglaise, allemande ou française.

Dans l'intérêt d'une publication rapide et d'une large diffusion les travaux publiés paraitront dans des cahiers individuels, diffusés séparément: 6 cahiers forment un volume.

En principe, seuls les manuscripts qui n'ont encore été publiés ni dans le pays d'origine ni à l'éntranger peuvent nous être soumis. L'auteur s'engage en outre à ne pas les publier ailleurs ultérieurement. Les auteurs recevront 25 exemplaires gratuits de leur publication.

Manuscripts should be addressed to / Manuskripte sind zu senden an / Envoyer les manuscrits à:

Prof. Dr. A. BRODAL, Universitetet i Oslo, Anatomisk Institutt, Karl Johans Gate 47 (Domus Media), Oslo 1 / Norwegen

Prof. W. HILD, Department of Anatomy, Medical Branch, The University of Texas, Galveston, Texas 77550/USA

Prof. Dr. J. van LIMBORGH, Universiteit van Amsterdam, Anatomisch-Embryologisch Laboratorium, Mauritskade 61, Amsterdam-O/Holland

Prof. Dr. R. ORTMANN, Anatomisches Institut der Universität, Lindenburg, D-5000 Köln-Lindenthal

Prof. Dr. T. H. SCHIEBLER, Anatomisches Institut der Universität, Koellikerstraße 6, D-8700 Würzburg

Prof. Dr. G. TÖNDURY, Direktion der Anatomie, Gloriastraße 19, CH-8006 Zürich/Schweiz

Prof. Dr. E. WOLFF, Lab. d'Embryologie Experimentale, College de France, 11 Place Marcelin Berthelot, F-75005 Paris/Frankreich

Advances in Anatomy, Embryology and Cell Biology
Ergebnisse der Anatomie und Entwicklungsgeschichte
Revues d'anatomie et de morphologie expérimentale

Vol. 55 · Fasc. 2

M. Dvořák

In Cooperation with J. Šťastná, S. Čech,
P. Trávník, D. Horký

The Differentiation
of Rat Ova
During Cleavage

With 62 Figures

Springer-Verlag Berlin Heidelberg New York 1978

Prof. MUDr. M. Dvořák, DrSc., MUDr. J. Šťastná, CSc., doc. MUDr. RNDr. S. Čech, CSc., MUDr. P. Trávník, CSc., MUDr. D. Horký, CSc., Department of Histology and Embryology, Medical Faculty of J. E. Prukyně University, Tř. Obránců míru 10, 662 43 Brno, ČSSR

ISBN-13: 978-3-540-08983-4   e-ISBN-13: 978-3-642-67047-3
DOI: 10.1007/978-3-642-67047-3

Library of Congress Cataloging in Publication Data. Dvořák, Milan., 1930–   Differentiation of rat ova during cleavage. (Advances in anatomy, embryology, and cell biology; v. 55, fasc. 2) Bibliography: p.   Includes index. 1. Ovum. 2. Cell differentiation. I. Šťastná, J. II. Title. III. Series. [DNLM: 1. Ovum Ultrastructure. 2. Ovum Cytology. 3. Rats Embryology. 4. Cell differentiation. W1 AD433K v. 55 fasc. 2 / QL965 D988d]. QL81.E67 vol. 55, fasc. 2 [QL965] 574.4'08s [599'. 3233] 78-13480

2121/3321-543210

Contents

1. Introduction and Notes on Materials and Methods

Milan Dvořák

The study of embryogenesis is increasingly concerned with the problems of the earliest stages of development. Mammalian ova, unlike those of lower vertebrates and invertebrates, have been largely neglected by researchers studying the problems of early embryogenesis, although they represent a unique object for investigation of the first differentiation processes. This is due to the fact that obtaining a great number of mammalian ova from the period of fertilization and, mainly, that of cleavage is connected with a number of technical difficulties which are further multiplied when the specimens are prepared for electron microscopy and ultracytochemistry. The possibilities for using segmenting mammalian ova in experimental studies are rather limited.

In our laboratory, we have been carrying out systematic research on the fine morphology of segmenting rat ova for years. We have, therefore, decided to present this comprehensive study of the ultrastructure and ultracytochemistry of segmenting rat ova based on our own results. The cleavage of the rat ovum takes place in a short period during the first 6 days of pregnancy and is accompanied by a number of morphologic, biochemical, and functional changes resulting in the formation of the blastocyst, i. e., an ovum capable of implantation. The study includes three main parts. The first, introductory part provides some information on the basic methods (details are given in the particular papers quoted). The second part contains detailed descriptive facts concerning the individual stages of rat ovum cleavage based on our results. Where necessary, this information is augmented by data from the literature which we had no opportunity of obtaining ourselves and which are important from the point of view of the problem studied. The descriptions are given in detail and separately for each stage of ovum development to enable the reader interested in only one stage to find all pertinent information in a concise form. The third part of the study includes discussions in which morphologic findings are compared with biochemical and functional ones. The references cover a broad scope of papers on the early embryogenesis of both mammals and other vertebrates and, in some cases, even some invertebrates.

Notes on Materials and Methods

The basis of this study are more than 2500 rat ova (Rattus norvegicus var. alba) obtained after fertilization and during cleavage at the one-, two-, four-, and eight-cell stages, the morula, and the early and late blastocysts. The ova were obtained by flushing the tuba uterina or the uterus and processed either for electron microscopy or ultracytochemistry.

For electron microscopic investigation, free ova were fixed in various ways, particularly by double fixation with 1.2% glutaraldehyde in cacodylate or phosphate buffer and with 1% OsO_4 in cacodylate or phosphate buffer. A simple fixation in 1% or 2% OsO_4 in phoshate buffer was used only exceptionally. The embedding medium was Durcupan ACM. Ultrathin sections were made by means of ultramicrotomes Tesla BS 490 or Ultrotome III LKB, stained with 1% aqueous uranyl acetate and lead citrate, and examined under the electron microscopes Tesla BS 500 or Tesla BS 613.

In preparing ova for electron microscopic investigation, part of the ova were studied ultracytochemically. In particular, the location of enzymes was studied: alkaline and acid phosphatases, nonspecific esterase, cholinesterase, succinate dehydrogenase, and endogenous peroxidase. Further more, the occurrence of polysaccharides, especially glycogen, and the glycocalyx were studied ultracytochemically. The ingestion of exogenous proteins of the horseradish peroxidase and microperoxidase type was also investigated.

Alkaline phosphatase was demonstrated by the method based on the Gomori technique with the incubation medium according to Mayahara et al. (1967) and acid phosphatase with the incubation medium according to Ericsson and Trump (1964–1965). To study the location of nonspecific esterase, the modified method according to Hanker et al. (1972) was used and for the proof of cholinesterase, the method according to Karnovsky (1964). Succinate dehydrogenase was studied by the method based on the reduction of ferricyanide with the incubation medium according to Ogawa et al. (1968). Endogenous peroxidase was demonstrated by the method according to Graham and Karnovsky (1966). Polysaccharides (glycogen) were ultracytochemically demonstrated by the method according to Thiéry (1967) and the glycocalyx by staining with ruthenium red according to Luft's (1971) method. Details about the methods used in the case of horseradish peroxidase are given in the paper by Dvořák and Trávník (1975) and in the case of microperoxidase in Dvořák and Trávník (1976). In addition to qualitative data, the present study is based on and augmented by quantitative data on the occurrence of cytoplasmic structures of the segmenting ova which were obtained by the morphometric method according to Weibel (Dvořák et al., 1977).

2. Submicroscopic Structure and Ultracytochemistry of Segmenting Rat Ova

Milan Dvořák and Jitka Šťastná

This section surveys submicroscopic and ultracytochemical changes which the rat ovum undergoes during the process of cleavage and which, to a certain extent, are the expression of the first transformation of a mammalian embryo in the process of differentiation or − taking into account the marked developmental lability even at the end of the cleavage process − are, at least, not insignificant for future differentiation.

Data are given successively in the individual stages of ovum development; the one-cell fertilized ovum, the two-cell ovum, the four-cell ovum, the eight-cell ovum, the morula, and the early and late blastocysts, as they were obtained in our laboratory. The stage of the unfertilized ovum has consciously not been included in the set of developmental stages of the rat ovum. We assume that it belongs, by nature of its topic, to the problems concerning fertilization, which are not dealt with in this study. In this respect, we refer the reader to the recent paper by Gwatkin (1976). As far as references are quoted in this section, they solely concern observations performed on rat ova. They are given only when they appear to be of extraordinary importance or when they differ from our findings, but they are only discussed in exceptional cases.

2.1. One-Cell Ovum

One-cell fertilized ova were obtained by flushing the tuba uterina at 10:00 a.m. - 2:00 p.m. on the 1st day of pregnancy. The average number of flushed ova in the morning hours was lower than in the afternoon, because it is easier to flush ova that have advanced from the abdominal orifice of the tuba uterina toward the uterus. According to our experience, the fertilized one-cell ovum can be obtained at the earliest at 7:00 a.m. on the 1st day of pregnancy in the favorable case. The earlier the ovum is taken, the more often the cells of the corona radiata occur in its neighborhood. At the period of sperm penetration, the ovum is in the metaphase of the meiotic division which, in the case of fertilization, is finished, and the separation of the second polar body into the perivitelline space takes place. The ovum then undergoes the so-called pronuclear phase (about 2 h after the penetration of the fertilizing sperm). At this stage, the ovum cell contains two pronuclei, a male one and a female one (Fig. 1). The male pronucleus has originated from the sperm proper and can already be identified in the light microscope, as the middle part of the sperm is usually found in its close vicinity. The description further refers to the fertilized one-cell ovum in the interphase at the stage of pronuclei, as no evidence has been obtained that a synkaryon is formed before the first segmenting division in rat ova. At this stage, the ovum is 47 - 50 μm, surrounded by the zona pellucida ($\sim$ 4.5 μm wide), and consists of the male and of the female pronuclei and cytoplasm (Fig. 1).

2.1.1. Nucleus

There are a male pronucleus and a female pronucleus in the one-cell ovum (Fig. 1). The volume of the two pronuclei increases, and the volumes of the pronuclei and the number of the nucleoli reach their maximum in about one-half at the pronuclear life span. The volume of the nucleoli increases more rapidly, so that the maximum is reached in about one-quarter of the pronuclear life span. The formation of the zygotic nucleus from the male and the female pronuclei is not known in mammals, perhaps only with the exception of monotremata (Austin, 1961).

Pronuclei are mostly situated in the cytoplasm excentrically. They reach a size of ca. 15 μm and are characterized by a very similar submicroscopic structure. Their shape is very irregular, owing to the indentation of their surface with deep invaginations and evaginations. Pronuclei of the rat ovum were studied in detail by Austin (1961) under the light microscope. He states that the male pronucleus is about 2.5 times larger than the female one; similar relations are valid for the number and the overall volume of nucleoli.

2.1.1.1. Nuclear Envelope

Pronuclei are limited by two membranes which border the perinuclear space, mostly 15 - 20 nm wide. In some places the perinuclear space is enormously widened due to numerous evaginations of the outer nuclear membrane; these dilatations often exceed 0.5 μm. In the perinuclear space, amorphous fine to flake-like material of low electron density is present. Dense bodies of variable size were often in that space (Fig. 2). They were limited by a membrane which, in some cases, was continuous with an inner membrane of the nuclear envelope. The material of these bodies is morphologically quite identical with that of the nucleoli. The presence of these bodies in the perinuclear space is probably the expression of one of the forms of extrusion of nucleolar material into the cytoplasm which occurs on a large scale at the stage of pronuclei (Szollosi, 1965a; Schuchner, 1970). Sotelo and Porter (1959) state that it is common for the nucleoli of female pronuclei to protrude beyond the surface of the nucleus.

Nuclear pores are irregularly distributed over the surface of the nucleus. Szollosi (1965a) states that the frequency of nuclear pores is very small. In places of their occurrence, the width of the perinuclear space is the same as is usually found in most

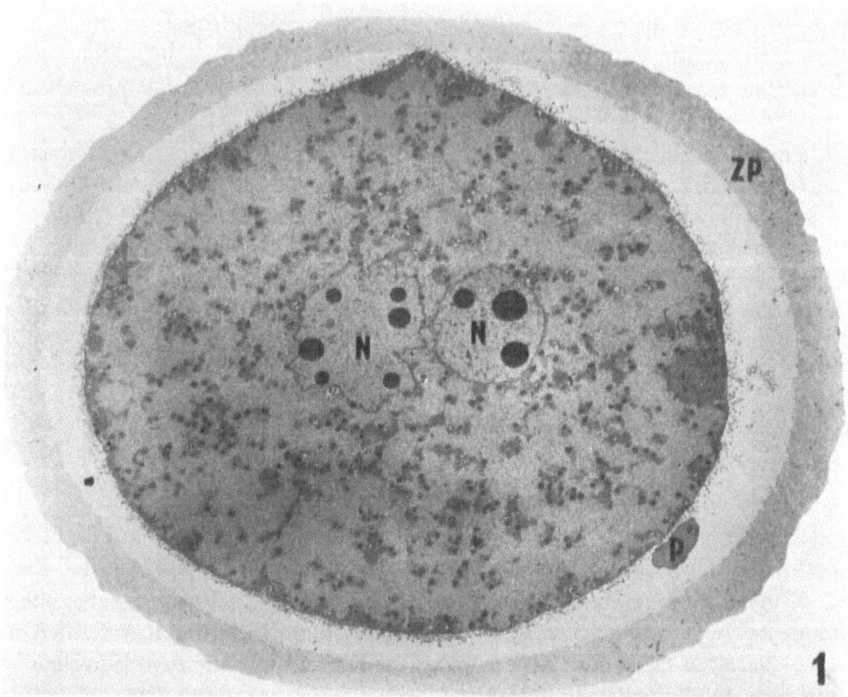

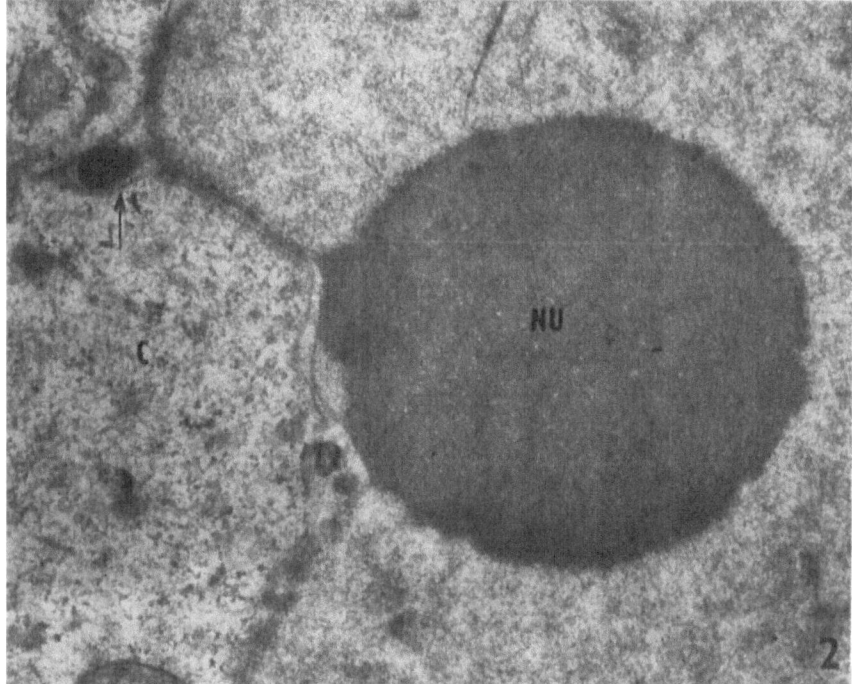

Figs. 1 and 2

Fig. 1. One-cell rat ovum. Two pronuclei (N) with distinct nucleoli; zona pellucida (ZP); polar body (P). Fixation glutaraldehyde and OsO₄; embedding medium Durcupan ACM; magnification X 1500

10

somatic cells (ca. 20 nm). Nuclear pores are, as a rule, filled with filamentous material with the same electron density as the nucleoli of the pronuclei.

Where there are dilatations of the perinuclear space, aggregations of vesicles of smooth endoplasmic reticulum are apparent in the adjacent ground cytoplasm. Similar vesicles in different stages of constriction have been observed in connection with the outer membrane of the nuclear envelope. On the nuclear envelope of the one-cell stage, some enzymes have been proved: acid phosphatase (Štastná, 1977a), alkaline phosphatase (Štastná, 1979), and organophosphate-sensitive nonspecific esterase (Trávník, 1977b, 1978).

2.1.1.2. Chromatin

Chromatin is distributed evenly in the nucleus (Figs. 1 and 3). Chromatin only forms minute condensations locally which, however, do not have the character of karyosomes. Mitosis with apparent chromosomes (Fig. 4) will not be discussed in this study.

2.1.1.3. Nucleolus

Nucleoli are prominent nuclear structures at the pronuclear stage (Fig. 3). Their volume is about 10% of the total nucleus volume in the rat at this stage (Austin, 1952). They are quite numerous; in our material, as many as ten nucleoli were found in one section of one pronucleus. Sotelo and Porter (1959) observed 12 - 17 and even more nucleoli in rat pronuclei. Austin (1961) states that the highest number of nucleoli found in the female pronucleus is 17 - 36. Frequently one large nucleolus of 6 μm diameter was observed in the pronucleus, and a great number of nucleoli of about 2 μm diameter and nucleoli or their fragments of considerably smaller size were observed. All nucleoli are compact and formed by the fibrilar component; occasionally nucleoli were also observed which seemed to start to form the granular component (Dvořák, 1974a).

The nucleoli are randomly distributed in the karyoplasm; part of them (nucleoli up to 2 μm diameter) are often located close to the inner membrane of the nuclear envelope (Fig. 2). According to the size and location, Szollosi (1965a) distinguishes three types of nucleoli at this stage; primary, secondary — already described by Austin (1952) — and tertiary nucleoli. According to Szollosi (1965a), tertiary nucleoli are small nucleoli or fragments of nucleolar material attached to the nuclear envelope. The nucleoli are primarily composed of protein material (Austin, 1952); in all three types, RNA was also proved (Szollosi, 1965a). In agreement with Szollosi (1965a) and Schuchner (1970), we found no signs of the presence of perinucleolar chromatin.

2.1.2. Cytoplasm

Cytoplasm is voluminous in the one-cell stage (Fig. 1). The cell organelles are distributed unevenly in the ground cytoplasm. They are principally concentrated in the

---

Fig. 2. One-cell rat ovum. Compact nucleolus (*NU*) attached to the nuclear envelope; evagination of the perinuclear space ($\rightarrow$) with a dark body; cytoplasm (*C*). Fixation glutaraldehyde and OsO_4; embedding medium Durcupan ACM; magnification X 32,000

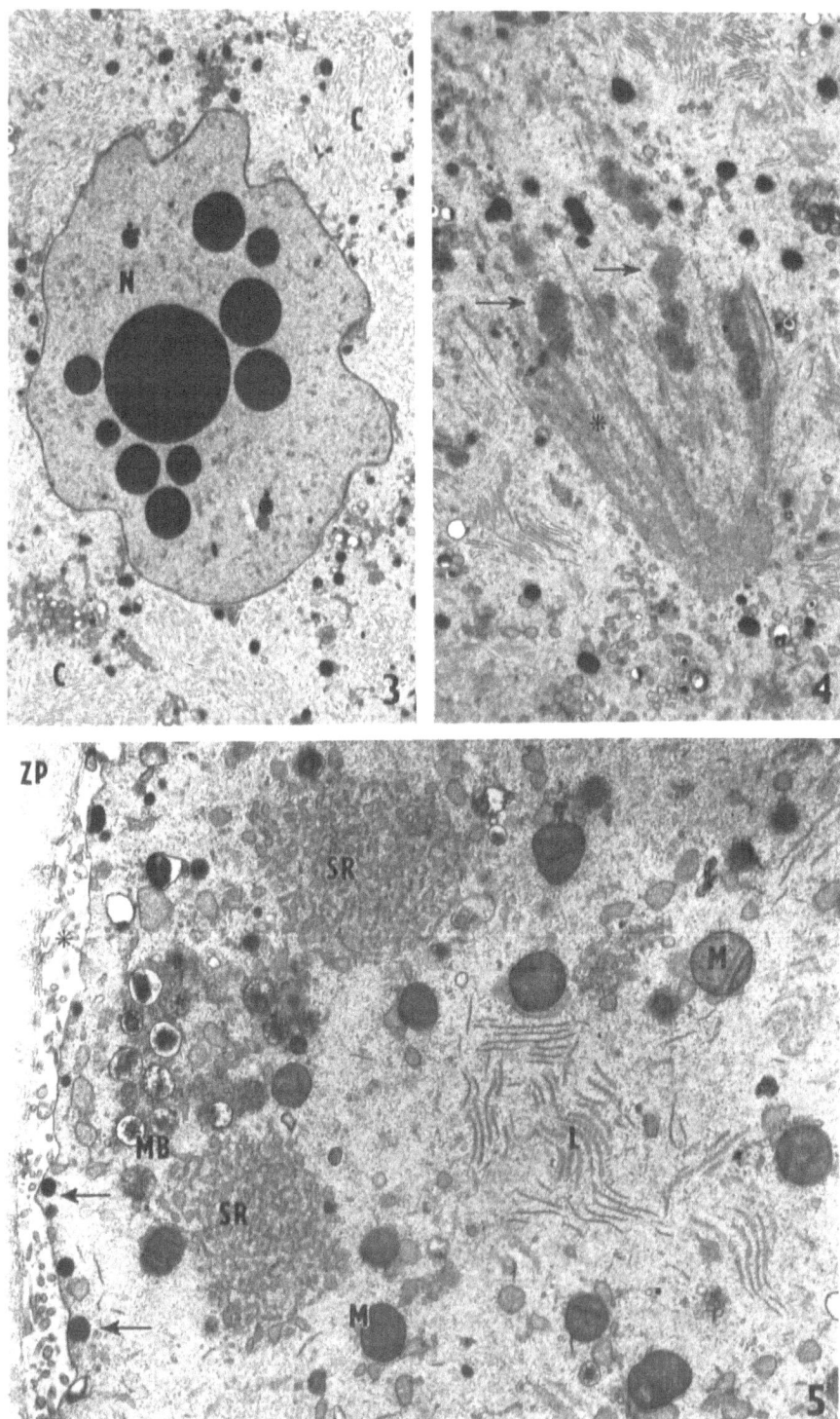

Figs. 3–5

cortical region (Fig. 5) and around the nucleus in small groups but are also found in other parts of the cytoplasm. Most of the ground cytoplasm is filled with lamellar structures and glycogen particles.

2.1.2.1. Mitochondria

Mitochondria which form 3.3% of the cytoplasmic volume (Dvořák et al. 1977), have an oval to round shape (Fig. 5). They are usually 0.5 - 0.7 μm large and are characterized by relatively uniform shape and size as well as by a simple inner structure. The matrix of the mitochondria has a higher than average electron density. Long cristae, frequently of typical arch-like course, project from the inner mitochondrial membrane into the matrix and reach as far as the opposite side of the mitochondria. In some cases, cristae of this arch-like course are arranged in parallel, two or three next to each other. Schlafke and Enders (1967) state that mitochondria are typically spheric and dense and have sparsely distributed septate or lamelliform cristae. A dense matrix makes the identification of intramitochondrial bodies and mitochondrial ribosomes difficult. Both have been discovered in mitochondria, intramitochondrial bodies occurring inconsistently and in small numbers. In the positive case, there was one body in one section of a mitochondrion.

In all mitochondria, the activity of succinate dehydrogenase was present (Šťastná, 1977b, 1978b). On mitochondrial cristae, the outer mitochondrial membrane, and partly also in the matrix, organophosphate-resistant esterase activity (Fig. 6) was found (Trávník, 1977b, 1978). Mitochondria are distributed irregularly in cytoplasm, but their occurrence is more frequent in the peripheral and central zones of the cytoplasm. From the point of view of their relation to other cytoplasmic structures, the most conspicuous and typical characteristic is the attachment of vesicles of smooth endoplasmic reticulum to their surface. It is not a rare phenomenon for an elongated cisterna of endoplasmic reticulum to pass from one mitochondrion to another, and mitochondria often occur in the region of complexes of multivesicular bodies.

2.1.2.2. Multivesicular Bodies

Multivesicular bodies are organelles of variable size and structure, characterized by the fact that the surface membrane limits numerous small vesicles located within the matrix (Fig. 7). The volume of multivesicular bodies is 2.1% of the total cytoplasmic volume (Dvořák et al., 1977), which means that they are a frequent cell organelle in the one-cell stage. Their size varies about 0.5 μm, and they are mostly somewhat smaller than mitochondria. Unusual variability in the appearance of multivesicular bodies is due to the variable content of elementary vesicles and the various amount of dense

Fig. 3. One-cell rat ovum. Nucleus (N) with ten nucleoli; cytoplasm (C). Fixation glutaraldehyde and OsO$_4$; embedding medium Durcupan ACM; magnification X 3,200

Fig. 4. One-cell rat ovum. Cell in mitosis; mitotic spindle (*); chromosomes (→). Fixation glutaraldehyde and OsO$_4$; embedding medium Durcupan ACM; magnification X 5,500

Fig. 5. One-cell rat ovum. Peripheral part of cytoplasm with mitochondria (M); smooth endoplasmic reticulum (SR); multivesicular bodies (MB); cortical granules (→); lamellar structures (L). In the perivitelline space (*) numerous microvilli; zona pellucida (ZP). Fixation glutaraldehyde and OsO$_4$; embedding medium Durcupan ACM; magnification X 14,000

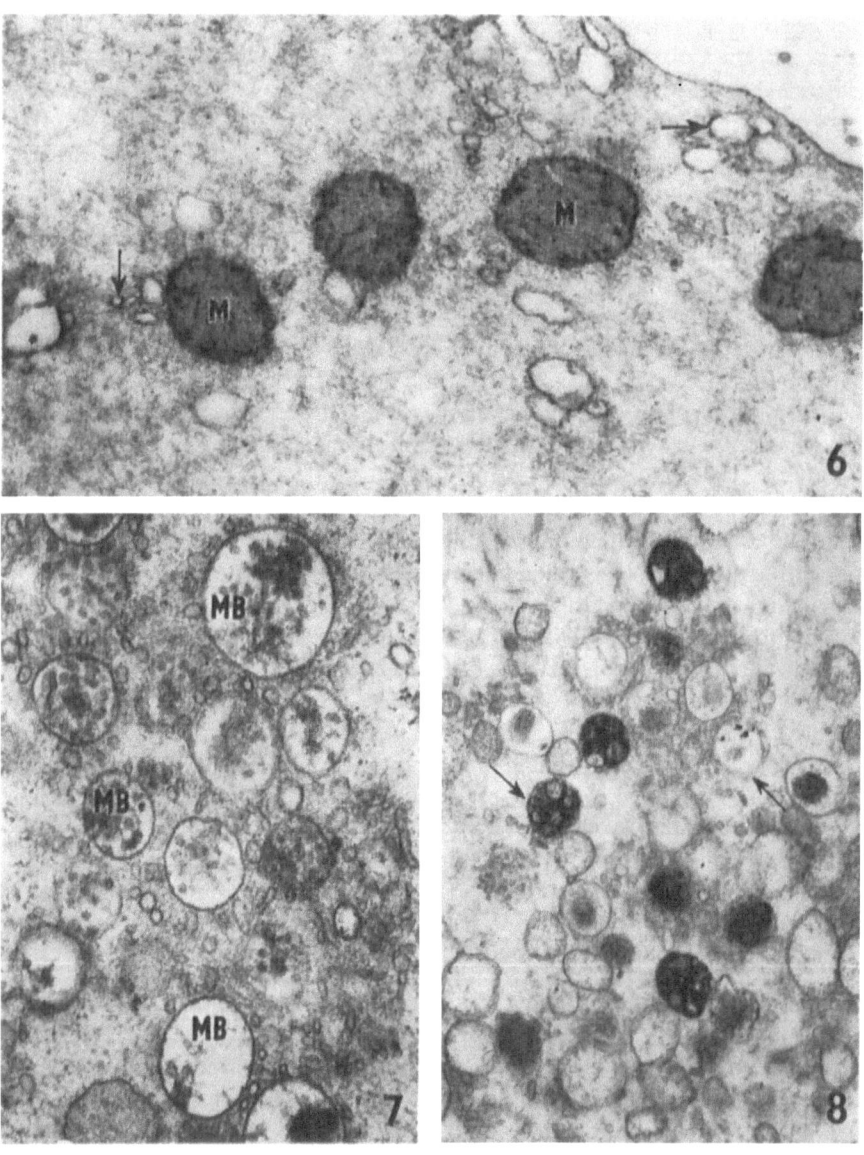

Fig. 6. One-cell rat ovum. Demonstration of organophosphate-resistant nonspecific esterase activity in mitochondria (*M*) and smooth vesicles (→). Fixation formaldehyde and OsO_4; embedding medium Durcupan ACM; nonstained ultrathin section; magnification X 39,000

Fig. 7. One-cell rat ovum. Numerous multivesicular bodies (*MB*) in various stages of development. Fixation glutaraldehyde and OsO_4; embedding medium Durcupan ACM; magnification X 33,000

Fig. 8. One-cell rat ovum. Demonstration of alkaline phosphatase activity in some multivesicular bodies (→). Fixation glutaraldehyde and OsO_4; embedding medium Durcupan ACM; nonstained ultrathin section; magnification X 24,000

amorphous matrix. Some bodies are almost completely filled with elementary vesicles; in other bodies, elementary vesicles only occur individually. The bodies without electron dense matrix and containing a low number of elementary vesicles appear to be light; bodies with a greater number of elementary vesicles and dense matrix appear to be dark. In some bodies, the matrix is concentrated approximately in their center; such bodies correspond to multivesicular bodies with a nucleoid, considered typical of rat ova by Sotelo and Porter (1959).

Multivesicular bodies are preferentially located in the peripheral zone of cytoplasm. There they are located in groups, often surrounded by regions of denser ground cytoplasm and frequently accompanied by smooth vesicles (Fig. 7). In the vicinity, cisternae of smooth endoplasmic reticulum also occur. In some multivesicular bodies, above all in those with dark matrix, alkaline (Fig. 8) and acid phosphatase activities are present (Šťastná, 1974a). No activity of nonspecific esterase has been found (Trávník, 1977b, 1978).

2.1.2.3. Autophagic Vacuoles

Autophagic vacuoles occur particularly at the periphery of the cell. In central regions, near the nuclei, they are located in aggregations of other cell organelles. Autophagic vacuoles are characterized by a size of up to 0.5 μm. As a rule, they are round in shape and contain smooth vesicles, membranes, and material morphologically similar to the ground cytoplasm, including glycogen particles. In some autophagic vacuoles, the components are at different stages of disintegration, so that in their morphologic analysis without cytochemical examination, it is difficult to draw a line between those structures and residual bodies.

As for topographic location, besides the above data, it is necessary to point out a frequent occurrence of autophagic vacuoles in the vicinity of multivesicular bodies. Autophagic vacuoles and residual bodies comprise 0.4% of the cytoplasmic volume (Dvořák et al., 1977). In autophagic vacuoles, the activity of acid phosphatase has been proved (Šťastná, 1977a), as well as that of alkaline phosphatase (Šťastná, 1979) and organophosphate-resistant esterase (Trávník, 1977b, 1978).

2.1.2.4. Residual Bodies

Residual bodies are those structures limited by membrane, the inner contents of which are characterized by the presence of particles whose original nature and detailed morphology can no longer be determined. Most often, it is intensely dense material in the form of an organized pseudomyeline figure or variously shaped structure of pseudolamellar character. Like in the case of autophagic vacuoles, the preferential place of their location is the periphery of the cytoplasm and the region near the nucleus. In some residual bodies, the activity of acid and alkaline phosphatases have been proved (Šťastná, 1974a, 1977a, 1979) as well as that of organophosphate-resistant esterase (Trávník, 1977b, 1978).

2.1.2.5. Granular Endoplasmic Reticulum

Granular endoplasmic reticulum has not been identified with certainty in the one-cell ovum. Schlafke and Enders (1967) did not find typical granular endoplasmic reticulum with associated ribosomes either.

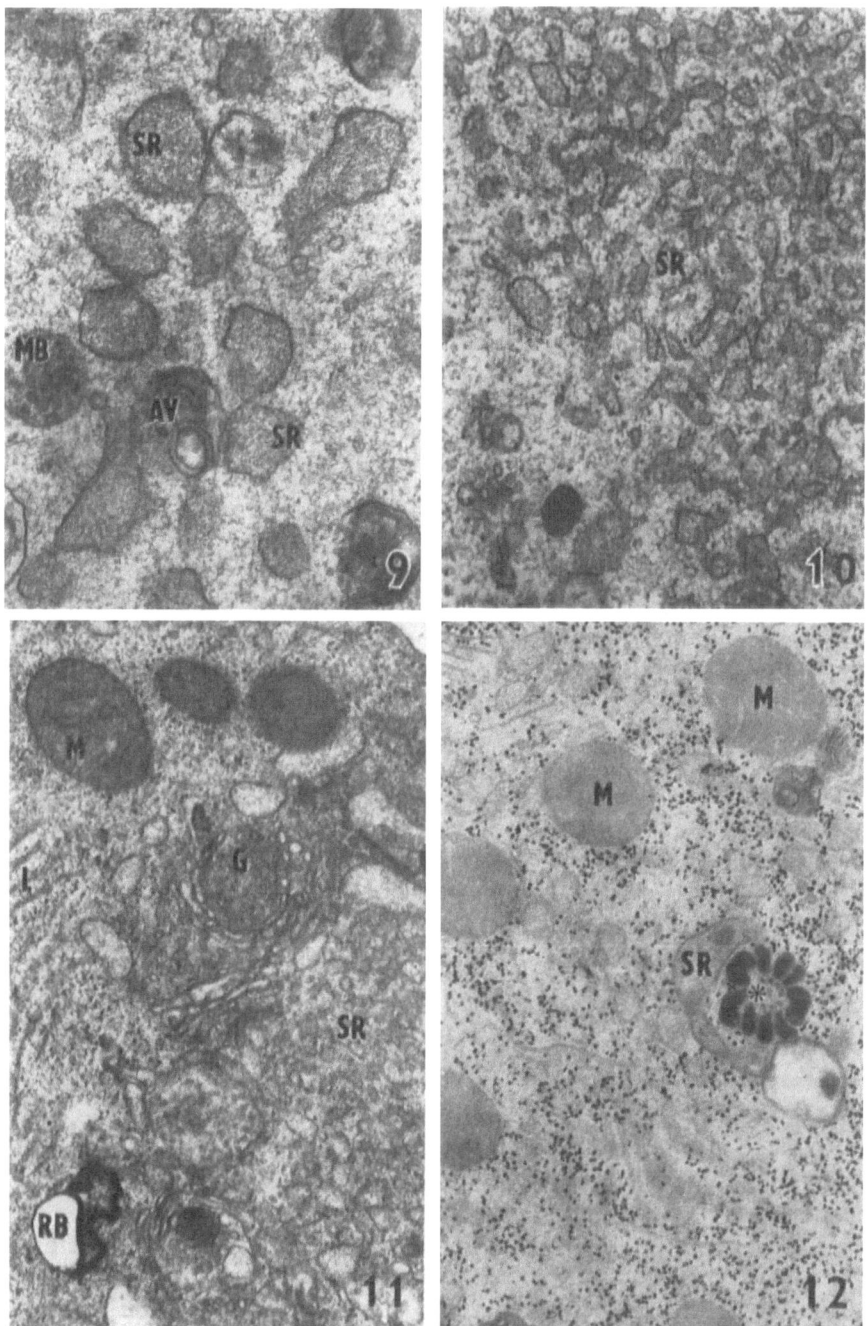

Figs. 9–12

Fig. 9. One-cell rat ovum. Large sacs of smooth endoplasmic reticulum (*SR*); multivesicular body (*MB*); autophagic vacuole (*AV*). Fixation glutaraldehyde and OsO$_4$; embedding medium Durcupan ACM; magnification X 32,000

2.1.2.6. Smooth Endoplasmic Reticulum

Smooth, vesicles belonging to this cell organelle occur in the one-cell ovum at a rate of about 1.9% of the cytoplasmic volume (Dvořák et al., 1977). In the electron microscopic picture, they appear in two forms: 1) as large sac-like structures (Fig. 9), sometimes flattened, 0.1 - 0.3 μm large and 2) as small vesicles, probably having the character of short tubules communicating with each other and grouped into areas in sections several square microns large (Fig. 10). Smooth endoplasmic reticulum mostly contains medium electron dense amorphous material. In some cases, a very dense core was observed in the center of the vesicle consisting of amorphous material that could not be distinguished in detail. Occasionally, the vesicles were empty under the electron microscope. Schlafke and Enders (1967) found large and rather elongated vesicles on the periphery of aggregations of smooth endoplasmic reticulum.

Large sacs have a frequent topographic relation toward mitochondria whose surface they sometimes approach and/or are in close contact with. Clusters of small vesicles are mostly located on the periphery of the cytoplasm. Both in complexes of smooth vesicles and in isolated vesicles distributed in the cytoplasm, the activities of acid phosphatase (Šťastná, 1974a, 1977a) and alkaline phosphatase (Šťastná, 1979) have been found. Vesicles of smooth endoplasmic reticulum in typical formations contain organophosphate-sensitive and also partly organophosphate-resistant esterases; isolated vesicles, particularly those in the peripheral regions of cytoplasm, reveal organophosphate-resistant esterase activity (Trávník, 1977b, 1978).

2.1.2.7. Ribosomes

Free ribosomes are difficult to distinguish in the cytoplasm of ova because their size and density resemble glycogen particles. Free ribosomes are present in the cytoplasm of the one-cell ovum in small quantities. Schlafke and Enders (1967) state that there are no rosettes or whorls of ribosomes present.

2.1.2.8. Golgi Complex

The Golgi complex is an inconspicuous structure in the one-cell ovum (Fig. 11). Comprises 0.4% of the cytoplasmic volume (Dvořák et al., 1977), and is only seldom noticed in ultrathin sections. It is formed by small Golgi fields composed of short, flattened cisternae, and in some cases, there are small vesicles in its vicinity whose origin might be derived from the Golgi complex. The Golgi complex is located partly in the vicinity of the nucleus, and, in some cases, also in cortical cytoplasm. Sotelo and Porter (1959) state that the distribution of the Golgi structures in the fertilized and segmenting rat ovum is more regular than in the preceding stages of ovum development. Cisternae of

---

Fig. 10. One-cell rat ovum. Aggregation of small vesicles, tubules of smooth endoplasmic reticulum (SR). Fixation glutaraldehyde and OsO$_4$; embedding medium Durcupan ACM; magnification X 32,000

Fig. 11. One-cell rat ovum. Golgi complex (G); mitochondria (M); residual body (RB); vesicles of smooth endoplasmic reticulum (SR); lamellar structures (L). Fixation glutaraldehyde and OsO$_4$; embedding medium Durcupan ACM; magnification X 34,000

Fig. 12. One-cell rat ovum. Demonstration of glycogen particles. Mitochondria (M); smooth endoplasmic reticulum (SR); middle piece of spermatozoon (*). Fixation glutaraldehyde and OsO$_4$; embedding medium Durcupan ACM; nonstained ultrathin section; magnification X 30,000

the Golgi complex are either empty or possess material of medium electron density. In all structural elements of the Golgi complex of the one-cell ovum, i. e., both in the cisternae and in the vesicles, the activities of acid phosphatase (Šťastná, 1974a, 1977a, 1978a) and alkaline phosphatase (Šťastná, 1979) have been proved.

2.1.2.9. Centrioles

Centrioles have not been found in the ova of our collection.

2.1.2.10. Lamellar Structures

Conspicuous structures, particularly from the quantitative point of view, are the lamellar structures, which account for 33.4% of the cytoplasmic volume (Dvořák et al., 1977). Lamellar structures appear in the cross-section like filaments ranging in length from 0.5 to 2.5 μm, depending on the site where the lamella was sectioned. Oblique sections show that lamellae are actually flat structures. In a section running parallel to its plane, the lamella clearly exhibits a grating pattern. The bars of this grating are about 34 nm and the width of the bar 14 nm. Lamellar structures are distributed throughout the whole cytoplasm, principally concentrated in the middle zone; they occupy large parts of cytoplasm free from any other cell organelles. As for their chemical nature they are protein in character, they do not contain RNA, and by the methods used in our experiments, no lipids, polysaccharides, or enzymes of the type of phosphatases, esterases, and dehydrogenases have been found (Dvořák et al., 1975).

2.1.2.11. Glycogen

Glycogen in the ova cytoplasm takes the form of particles (β-type) reaching a size of 30 nm (Fig. 12). In areas of aggregations of lamellar structures, small smooth vesicles, and multivesicular bodies, the number of glycogen particles is considerably lower than in other regions. Inside cell organelles, glycogen particles rarely occur and were only observed in some residual bodies. Aggregations of glycogen particles are situated close to some multivesicular bodies; sometimes, even the onset of separation of glycogen aggregations from the surrounding cytoplasm took place at the site of those organelles. Individual glycogen particles were even found in the microvilli of the ovum surface.

2.1.2.12. Lipid Droplets

Lipids in the form of droplets were only found occasionally, comprising 0.1% of the cytoplasmic volume (Dvořák et al., 1977). They are individual round particles not limited by membrane. Owing to their small quantity, it is not possible to determine the predilection of their location.

2.1.2.13. Cell Membrane

Cell membrane surrounding the cytoplasm of the ovum extends in the form of short and numerous microvilli into the perivitelline space. Among the microvilli, signs of the formation of pinocytotic vesicles can be observed locally. The distribution of microvilli on the surface of the ovum is regular. Sotelo and Porter (1959) state that after fertilization the microvilli disappear and the cell membrane appears to disrupt in many places. Cytoplasmic material emerges from the multiple openings in the membrane.

Restitutio ad integrum of the cell membrane was not observed until the first cleavage division. The findings of the above authors must be attributed to the technique of material processing (embedding in methacrylates); some "apparent" openings could be places of release of cortical granules into the perivitelline space. The glycocalyx appears after staining with ruthenium red as a continuous dark line 20 nm wide, copying the course of the cell membrane (Dvořák, 1977). On the cell membrane of one-cell ovum, the alkaline phosphatase actitvity was found infrequently (Šťastná, 1979).

2.1.2.14. Cortical Granules

Cortical granules are intensely electron dense ovoid structures located below the cell membrane at different distances from each other. They are absolutely characteristic of unfertilized ova; in fertilization, most of those granules disappear due to the so-called cortical reaction. In spite of that, even in the fertilized one-cell ovum, individual intact cortical granules sometimes remain in the cytoplasm. The same finding was reported by Schlafke and Enders (1967).

2.1.3. Zona Pellucida

The zona pellucida surrounds the one-cell ovum so that only a narrow perivitelline space remains between its surface and the inner surface of the zona pellucida. The width of the zona pellucida is relatively constant, about 4.5 μm. Structurally, in the electron microscopic picture, it appears to consist of filaments forming fine webs and granular material. The outer part of the zona pellucida is more sparsely arranged than the inner part.

2.2. Two-Cell Ovum

The two-cell ova were obtained at 7:00 a.m. - 9:00 a.m. on the 2nd day of pregnancy. Blastomeres are equivalent from the morphological point of view (Fig. 13). The size of the ovum remains approximately the same as at the one-cell stage, which means that the blastomeres thus developed have a substantially smaller size than the zygote. In the perivitelline space, two polar bodies are located as a rule. The blastomeres, which are ovoid with flattening at the sites of their contact, are 50 μm long and 25 μm wide. The ovum is surrounded by the zone pellucida, and each of the blastomeres consists of the nucleus and cytoplasm.

2.2.1. Nucleus

The nucleus is situated approximately in the center of the blastomere and is 11.5 μm large (Figs. 13 - 15). The shape of the nucleus is slightly oval with a surface exhibiting shallow irregularities as larger invaginations and evaginations. Structurally, the nucleus consists of the nuclear envelope, chromatin, and nucleoli.

2.2.1.1. Nuclear Envelope

The nuclear envelope is formed by two membranes separated by the perinuclear space. The nuclear envelope is not smooth, and its surface manifests irregularities which can

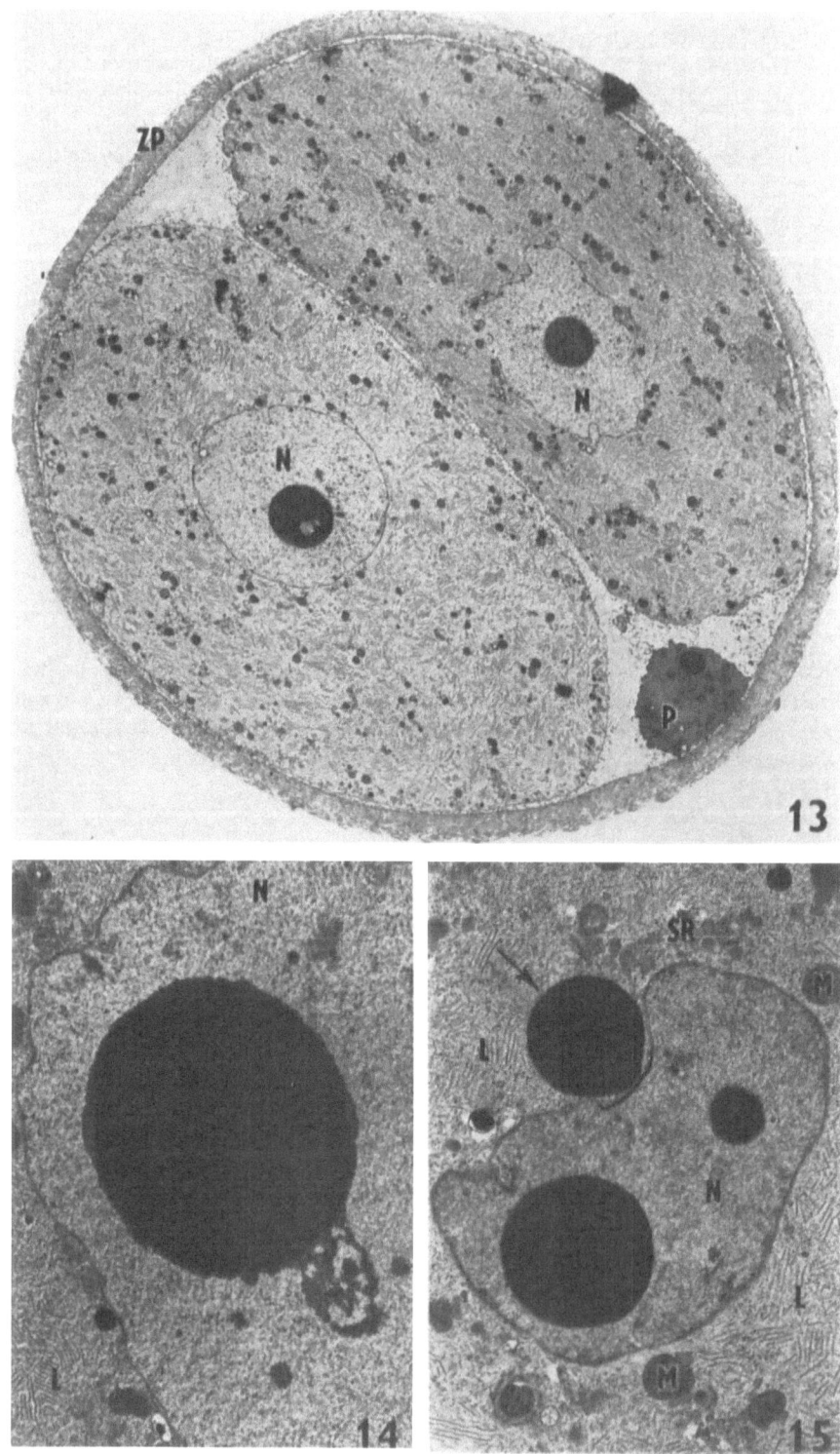

Figs. 13–15

be characterized as invaginations and evaginations of various depth (Figs. 14 and 15). Deep invaginations are always narrower than shallow ones. The invaginations inside the nucleus are not due to folds of the nuclear envelope but rather, in some cases, to the invagination of only the inner nuclear membrane, in which case they have a finger-like appearance, and the surface of the nucleus limited by the outer membrane remains smooth.

The perinuclear space is mostly narrow, measuring 15 - 20 nm. In places it is, however, slightly broadened. The broadened portions lie between two neighboring pores. In addition, in individual cases, major extensions of the perinuclear space are perceptible due to the invagination of the inner nuclear membrane, in which case the outer nuclear membrane is not evaginated. Medium dense amorphous material is present in the perinuclear space. Nuclear pores are numerous. Inside the pore, intensely electron dense material is regularly located extending to both sides of the pore. The distribution of the pores is irregular. Often, the nuclear pores are situated in groups, leaving major parts of the nuclear envelope pore-free.

The tiny condensations of chromatin near the inner face of the inner nuclear membrane are extremely rare. Occasionally, round, intensely electron dense structures are situated at the membrane or close to it. Their size is about 0.2 μm and structurally they correspond to chromatin condensation. From the inner membrane, structures corresponding to intranuclear annulate lamellae emerge inside the nucleus. In one case, a special feature of the nuclear envelope was observed at a site where there had been an extrusion of a nucleolus (Fig. 15). In that region, there is a deep invagination in the nucleus, and here the individual components of the nuclear envelope cannot be distinguished, the nucleus being bordered by a dense layer (plate) 30 nm thick. Close to the outer nuclear membrane, there are sacs with intensely electron dense amorphous contents, and empty smooth sacs. In the perinuclear space, acid phosphatase (Šťastná, 1977a), alkaline phosphatase (Šťastná, 1979), and organophosphate-sensitive esterase activity (Trávník, 1977b, 1978) has been proved.

2.2.1.2. Chromatin

Chromatin is distributed evenly in the nucleus, which gives the nucleus a light appearance in the electron microscopic picture (Fig. 13). In comparison with the nucleus of the one-cell stage, it is necessary to draw attention to a far more frequent occurrence of tiny flake-like aggregations of chromatin, which are 50 - 500 nm in size. In the karyoplasm, structures consisting of dark granules about 150 nm in size are sometimes also present.

Fig. 13. Two-cell rat ovum. Two blastomeres with nuclei (N) and distinct nucleoli; zona pellucida (ZP); polar body (P). Fixation glutaraldehyde and OsO$_4$; embedding medium Durcupan ACM; magnification X 2,000

Fig. 14. Two-cell rat ovum. A part of nucleus (N) with reticulated nucleolus, surrounded by cytoplasm filled with lamellar structures (L). Fixation glutaraldehyde and OsO$_4$; embedding medium Durcupan ACM; magnification X 8,000

Fig. 15. Two-cell rat ovum. Nucleus (N) with compact nucleoli, one of which extrudes into the cytoplasm (→). Mitochondria (M); smooth endoplasmic reticulum (SR); lamellar structures (L). Fixation glutaraldehyde and OsO$_4$; embedding medium Durcupan ACM; magnification X 8,000. From Dvořák, M., Scripta med. 47, 497–502 (1974)

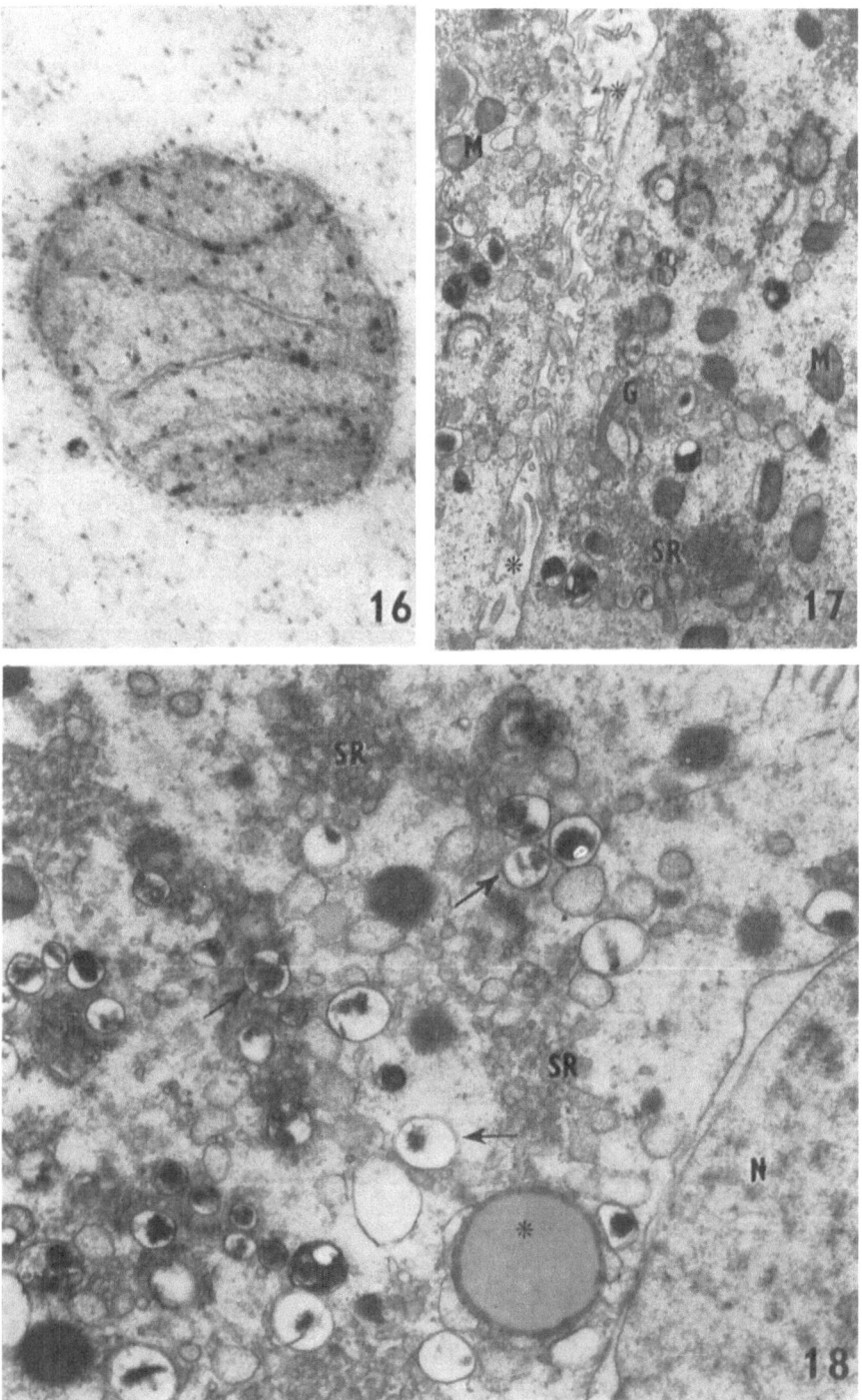

Figs. 16–18

Fig. 16. Two-cell rat ovum. Demonstration of succinate dehydrogenase activity in the mitochondrion, which is characterized by long arch-like cristae. Fixation OsO$_4$; embedding medium Durcupan ACM; nonstained ultrathin section; magnification X 65,000

2.2.1.3. Nucleolus

In the two-cell stage, nucleoli occur in smaller numbers than in the one-cell stage. In ultrathin sections, one to three nucleoli can be observed, most often only one to two nucleoli (Fig. 13). The nucleoli are sometimes located on the periphery of the nucleus, their size being as large as 5 μm. Most of the nucleoli (Fig. 15) belong to the group of compact nucleoli (62%), but nucleoli also occur in two-cell ova with first signs of the development of nucleolonema on the periphery (Fig. 14), i. e., nucleoli differentiating into the reticular type (about 32%). In some cases, extrusion of nucleoli of the compact type into the cytoplasm was observed (Fig. 15). Quite rarely, ring-shaped nucleoli could be seen. Schuchner (1970) states that ring-shaped nucleoli have been observed at that stage and that the pars granulosa is absent.

Szollosi (1966) and Schuchner (1970), however, only noticed nucleoli of the compact type. From our observations as well as from some published data (Szollosi, 1971), it seems probable that the occurrence and appearance of nucleoli depend to a large extent on the age of the two-cell ovum. In ova obtained soon after the first mitosis, only compact nucleoli were observed; in those taken later during the interphase, the occurrence of nucleoli with the developing pars granulosa increased. Ring-shaped nucleoli were observed in only about 6% of our cases of the two-cell ovum nucleus. This finding does not quite agree with earlier findings by Sotelo and Porter (1959) who consider ring-shaped nucleoli characteristic of rat ova. The presence of this type of nucleoli in the two-cell ovum was also noted by Schuchner (1970), who, however, does not give any details about the frequency of their occurrence. In agreement with other authors (Sotelo and Porter, 1959; Izquierdo and Vial, 1962; Mazanec, 1965; Szollosi, 1966, 1971), we also observed compact nucleoli located randomly in the cytoplasm (Fig. 15). Izquierdo and Vial (1962) think that some of these nucleoli extrude into the cytoplasm through openings in the nuclear envelope. The other authors assume that these nucleoli are those which remained in the cytoplasm during mitosis.

2.2.2. Cytoplasm

The cytoplasm surrounds the nucleus and is limited on the surface by the cell membrane with numerous microvilli. Cell organelles are located rather irregularly among lamellar structures, their concentration being higher in the vicinity of the nucleus and on the cell periphery.

2.2.2.1. Mitochondria

Mitochondria form about 4% of the cytoplasmic volume (Dvořák et al., 1977). Their shape is always slightly oval (Fig. 16); rod-shaped mitochondria are not present in the

---

Fig. 17. Two-cell rat ovum. Contact of two blastomeres with wide intercellular space (∗) filled with microvilli. In the cytoplasm, Golgi complex (G); mitochondria (M); vesicles of smooth endoplasmic reticulum (SR). Fixation glutaraldehyde and OsO₄; embedding medium Durcupan ACM; magnification X 13,000

Fig. 18. Two-cell rat ovum. Part of a blastomere with nucleus (N) and cytoplasm containing multivesicular bodies (→) in different developmental stages, vesicles, and sacs of smooth endoplasmic reticulum (SR); lipid droplet (∗). Fixation glutaraldehyde and OsO₄; embedding medium Durcupan ACM; magnification X 19,000

two-cell stage. Their size varies from about 0.5 to 0.7 μm. Mitochondria are characterized by a small number of cristae which in some cases have arch-like courses. The matrix of mitochondria is electron dense. Mitochondria are distributed in cytoplasm in groups of two to five. They are closely related to smooth endoplasmic reticulum whose cisternae often touch the surface of the mitochondrion (Fig. 17). On the cristae of mitochondria and the outer membrane, succinate dehydrogenase (Fig. 16) (Šťastná, 1977b, 1978b) and organophosphate-resistant esterase activity has been found (Trávník, 1977b, 1978).

2.2.2.2. Multivesicular Bodies

Multivesicular bodies comprise 1.3% of the cytoplasmic volume (Dvořák et al., 1977). They are sac-like structures, 0.25 - 0.5 μm large, surrounded by a membrane, and their inner contents are fairly variable (Fig. 18). Multivesicular bodies may be distinguished in two basic forms. The first one involves light bodies characterized by the presence of a small number of vesicles (size $\sim$ 25 nm) with medium dense contents inside the body which is otherwise, particularly on the periphery, empty. The second type are dark multivesicular bodies filled with electron dense material; elementary vesicles inside them appear light (size ca. 50 nm and more). The multivesicular bodies most often occur with other organelles, such as smooth vesicles, smooth tubules, and mitochondria in regions of cytoplasm near the nucleus and cell membrane. Some of the multivesicular bodies showed acid phosphatase (Šťastná, 1974a, 1977a) and alkaline phosphatase activity (Šťastná, 1979).

2.2.2.3. Autophagic Vacuoles

In the two-cell ovum, autophagic vacuoles account for 0.7% of the cytoplasmic volume (Dvořák et al., 1977) and are bodies of 0.5 μm large, but may also be smaller. They are enclosed by a membrane and contain all membrane structures mentioned above. In some, particularly small autophagic vacuoles, glycogen particles occur. The contents of the vacuoles are characterized by a different degree of degradation. Autophagic vacuoles are always present in the cytoplasmic regions where other cell organelles aggregate. Acid phosphatase (Šťastná, 1977a), alkaline phosphatase (Šťastná, 1979), and organophosphate-resistant esterase activities (Trávník, 1977b, 1978) have been observed.

2.2.2.4. Residual Bodies

In this stage, residual bodies are present in a small quantity only. They are enclosed by a membrane and in some cases, their contents have the appearance of pseudomyelins, while in other cases they appear as intensely dense material without any structure. In the majority of the residual bodies, acid phosphatase (Šťastná, 1974a, 1977a) and alkaline phosphatase (Šťastná, 1979) activity has been found and in a few instances organophosphate-resistant esterase (Trávník, 1977b, 1978) activity has been observed.

2.2.2.5. Smooth Endoplasmic Reticulum

Smooth endoplasmic reticulum takes up 1.6% of the cytoplasmic volume (Dvořák et al., 1977). It is composed of sac-like structures, 0.1 - 0.5 μm in size, mostly filled with amorphous to flake-like material of medium electron density. They occur in groups or are singly distributed. Often, they are attached to mitochondria, in which case they are flattened, partly encircling the mitochondrion. Another form of smooth endoplasmic reticulum are small vesicles and/or tubules, up to 100 nm in size, with medium electron dense contents. These tiny tubules are concentrated in large fields near the nucleus or on the periphery of the cytoplasm, and in cross-section they occupy an area of 0.5 μm^2 and more. In the smooth endoplasmic reticulum, both alkaline and acid phosphatases were regularly present (Šťastná, 1974a, 1977a, 1979) as well as nonspecific organophosphate-sensitive esterase and in part of the vesicles also organophosphate-resistant esterase (Trávník, 1977b, 1978).

2.2.2.6. Granular Endoplasmic Reticulum

Granular endoplasmic reticulum has not been found in our material.

2.2.2.7. Ribosomes

Ribosomes occur only occasionally. They are mostly distributed separately. Polysomes were found in the vicinity of the mitotic spindle. Szollosi (1966) and Schlafke and Enders (1967) also observed the presence of only a small quantity of polysomes.

2.2.2.8. Golgi Complex

The Golgi complex is present in the two-cell ovum only at the rate of 0.1% of the cytoplasmic volume (Dvořák et al., 1977). The Golgi complex is more often located on the periphery of the cytoplasm. Schlafke and Enders (1973) observed that the Golgi elements were located on the periphery of the membrane complex consisting of smooth endoplasmic reticulum and differently joined membrane structures. It is formed by three to four flattened cisternae 0.5 - 1 μm long, at whose ends and also in the concavity of the Golgi complex small vesicles 100 nm in size are situated. The third component is formed by several large smooth sacs which cannot be distinguished from the structures of smooth endoplasmic reticulum. In the vicinity of the Golgi complex, multivesicular and residual bodies, autophagic vacuoles and vesicles of smooth endoplasmic reticulum are often located. At the stage of the two-cell ovum, acid phosphatase (Šťastná, 1974a, 1977a, 1978a) and alkaline phosphatase activity (Šťastná, 1979) has been regularly proved in the Golgi complex and in isolated cases also the activity of organophosphate-sensitive esterase (Trávník, 1977b, 1978).

2.2.2.9. Centrioles

Typical centrioles have not been found in our material. During mitosis, a formation was observed in the region of the apex of the mitotic spindle, morphologically resembling a centriole. Studies on maturing rat eggs support the hypothesis that centrioles in general may be absent in early stages of mammalian development (Szollosi, 1975b).

2.2.2.10. Lamellar Structures

Lamellar structures take up 34.7% of the cytoplasmic volume (Dvořák, et al., 1977). They extend in different directions in the cytoplasm, being aggregated into groups formed by a great number of lamellae (Fig. 19). They exclude only a narrow zone near

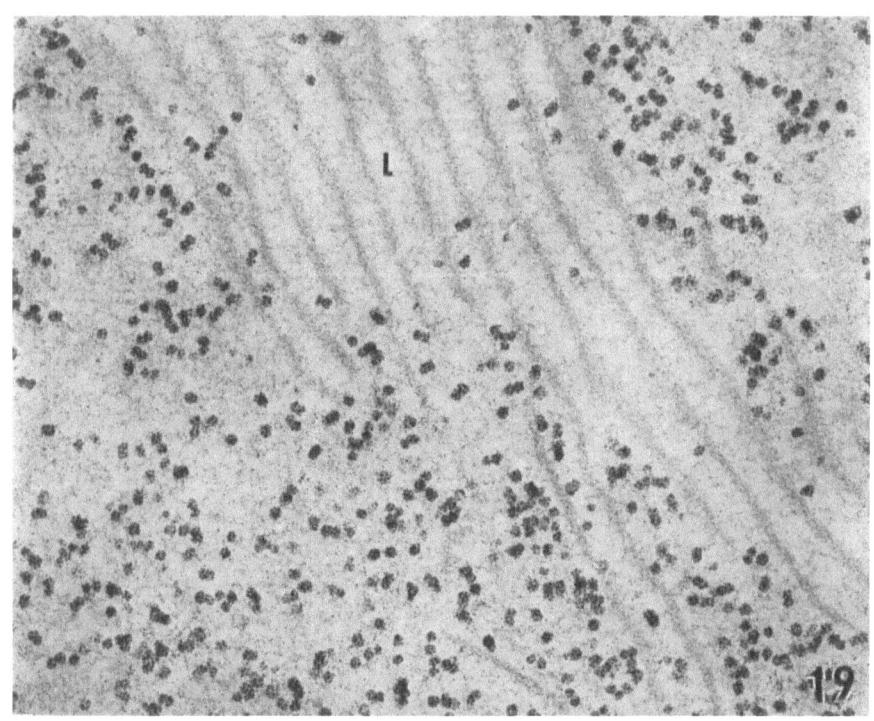

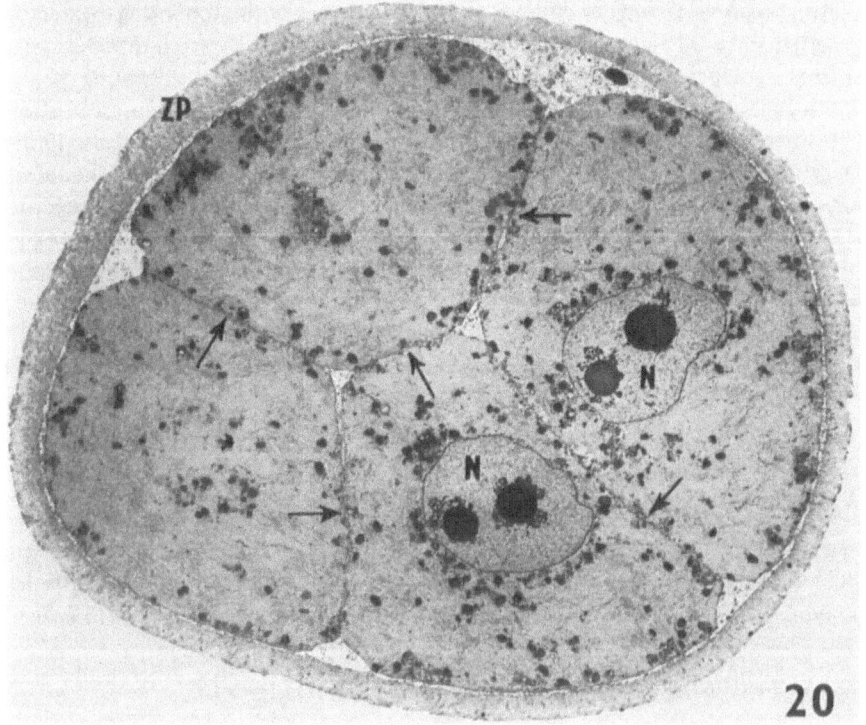

Figs. 19 and 20

the cell membrane but reach close to the nuclear envelope. Lamellar structures occur only rarely in regions occupied by cell organelles. No degradation signs of them are visible.

2.2.2.11. Glycogen

In the electron microscopic picture, glycogen has the form of particles (β-type) scattered in the ground cytoplasm, but they only occur rarely among lamellar structures (Fig. 19). This basic pattern is completed by massive aggregation of glycogen particles in clusters which are usually enclosed in major smooth sacs, corresponding to autophagic vacuoles (Čech, 1977a, b, c). The average size of those sacs is 0.4 - 0.6 μm. Smooth sacs with glycogen are located among cell organelles, frequently showing close topographic relations to mitochondria. They were even found in the perinuclear space. Very often, aggregations of glycogen particles were found in various degrees of enclosure by smooth sacs, in which case they had a horseshoe-like appearance (Čech, 1977a). Larger clusters of glycogen particles also occur in the vicinity of and/or even in some multivesicular bodies. As for the frequency of those observations, one can state that they did not differ significantly from those of one-cell ova. The same can be stated for the presence of single glycogen particles in some residual bodies (Čech, 1977a).

2.2.2.12. Lipid Droplets

Lipid particles were only found occasionally in the two-cell stage (Fig. 18); thus, they cannot be evaluated quantitatively. In one case observed, it was a lipid vacuole (1 μm in size) enclosed by a membrane in vicinity of the nucleus.

2.2.2.13. Cell Membrane

The cell membrane projects into short microvilli both on the free surfaces of the blastomeres and on the surfaces facing the intercellular space. The contact between the blastomeres is rather loose (Fig. 17). The intercellular gap is as wide as 0.3 μm. On the surface of adjacent blastomeres, there are numerous microvilli, often touching each other. The intercellular space is completely filled with material staining, like the glycocalyx, with ruthenium red. There are no junctional structures between the blastomeres (Dvořák, 1977). On the free surface of the blastomeres, microvilli are somewhat more sparsely distributed. The perivitelline space there is 0.3 μm wide and some microvilli even touch the zona pellucida.

2.2.2.14. Cortical Granules

Cortical granules have not been found in the two-cell stage. Schlafke and Enders (1967) found only isolated cortical granules in such ova.

Fig. 19. Two-cell rat ovum. Demonstration of glycogen particles. Glycogen is absent in areas filled with lamellar structures (*L*). Fixation glutaraldehyde and OsO$_4$; embedding medium Durcupan ACM; nonstained ultrathin section; magnification X 80,000

Fig. 20. Four-cell rat ovum. Contacts of blastomeres ($\rightarrow$); nuclei (*N*); zona pellucida (*ZP*). Fixation glutaraldehyde and OsO$_4$; embedding medium Durcupan ACM; magnification X 2,000

2.2.3. Zona Pellucida

The zona pellucida is about 3 μm wide and its morphologic picture is the same as in the preceding stage. It is separated from the cell membrane of the ovum by a narrow perivitelline space 0.3 μm wide. After exposure to the action of ferritin (Schlafke and Enders, 1973) for about 10 min. ferritin was only found in the zona pellucida; horseradish peroxidase was even found on the surface of the cell membrane after this exposure time.

2.3. Four-Cell Ovum

The four-cell ova (Fig. 20) were obtained at 7:00 a.m. - 9:00 a.m. on the 3rd day of pregnancy. The size of the individual blastomeres is essentially the same, their diameters being 23 μm; the size of the ovum is about 50 μm. The blastomeres are of polygonal shape. The contact surfaces of the blastomeres are flattened, and the intercellular spaces are narrow. They are dilated, as a rule, only at the site of contact of three or four blastomeres. The perivitelline space is also narrow and only somewhat broadened at the site of contact of the adjacent blastomeres. In the perivitelline space, one or two degenerating polar bodies can be observed. The individual blastomeres consist of the nucleus and cytoplasm.

2.3.1. Nucleus

The nucleus is located approximately in the center of the cell (Fig. 20); it is of slightly oval shape, its size being 10 μm. Its surface shows only shallow irregularities due to the invagination of the nuclear envelope. Structurally, the nucleus consists of the nuclear envelope, chromatin, nucleoli, and inconsistently present intranuclear annulate lamellae.

2.3.1.1. Nuclear Envelope

The perinuclear space of the nuclear envelope is narrow, 15 - 20 nm wide (Fig. 21) and only slightly dilated in places. In individual cases, the inner membrane projects into the nucleus in the form of invaginations corresponding to intranuclear annulate lamellae. The outer nuclear membrane also shows evaginations which are wider and shorter. As has been stated before, the nuclear surface is, on the whole, smooth and only in places are shallow foldings of the nuclear envelope evident. Mazanec and Dvořák (1963) also state that at this stage invaginations of the nuclear envelope disappear gradually. The perinuclear space is sometimes bridged by nuclear pores which in some places are numerous in section (up to ten pores per 1 μm) but in other places distributed at greater distances (five pores per 1 μm). The regions of nuclear pores are electron dense and in this respect more conspicuous than the adjacent part of the cytoplasm. In the perinuclear space, organophosphate-sensitive esterase activity has been proved (Trávník, 1977b, 1978) and sometimes also acid phosphatase (Šťastná, 1977a) and alkaline phosphatase activity (Šťastná, 1979).

2.3.1.2. Chromatin

Chromatin, distributed evenly in the nucleus, occasionally shows small condensations and, near the inner nuclear membrane, a continuous dark rim interrupted only at the sites of nuclear pores.

2.3.1.3. Nucleolus

In the nuclei of the individual blastomeres, one to two nucleoli are present (Figs. 20 and 22). These are only seldom larger than 3 μm. They may be characterized as a reticularly transforming type of compact nucleoli. Their basic component is the compact part of the nucleolus on whose periphery nucleolonemata start to form or can even be fairly developed. In some cases, when two nucleoli are located close to each other, nucleolonemata join. It has often been observed that an extensive formation of the pars granulosa takes place particularly on that side of the nucleolus facing the nuclear envelope. In evaluating the individual types of nucleoli occurring in the four-cell stage, a ring-shaped nucleolus was rarely observed. Reticular rearrangement of the nucleolus at the four-cell stage was also observed by Szollosi (1966) and Schuchner (1970).

2.3.2. Cytoplasm

In the cytoplasm, cell organelles are distributed unevenly, in places they are concentrated into groups, particularly at the periphery of blastomeres and perinuclearly (Fig. 20). For the most part, the cytoplasm is occupied by lamellar structures.

2.3.2.1. Mitochondria

Mitochondria take up 4% of the cytoplasmic volume (Dvořák et al., 1977). Their basic shape is slightly oval (Fig. 21) but also circular and short rod-like profiles can be encountered. Therefore, even the average size of mitochondria varies about 0.5 μm and more. Mitochondria still have dense matrix, and mitochondrial cristae are not numerous. They often have an arch-like course and bridge the whole breadth of the mitochondrion. Frequent lamellar-like arrangements of several cristae in one pole of the mitochondrion have been observed. Due to the high density of mitochondrial matrix, the distinction between intramitochondrial bodies and mitochondrial ribosomes is not possible. Mitochondria are located in the cytoplasm separately or in small groups. Smooth endoplasmic reticulum has a close relation to mitochondria, and its flattened vesicles are closely attached to the surfaces of mitochondria. In their vicinity, multivesicular bodies and lysosomes can be located.

On mitochondrial cristae, succinate dehydrogenase (Šťastná, 1977b, 1978b) and organophosphate-resistant esterase activity (Trávník, 1977b, 1978) has been proved.

2.3.2.2. Multivesicular Bodies

Multivesicular bodies form 0.4% of the cytoplasmic volume (Dvořák et al., 1977). Their number is significantly lower than in the preceding stages. They occur in two forms. Light multivesicular bodies contain a small number of tiny vesicles, about 30 nm in size, some of which however, reach a size of 100 nm. The contents of the multivesicular body around the vesicles are formed by electron light finely granular mate-

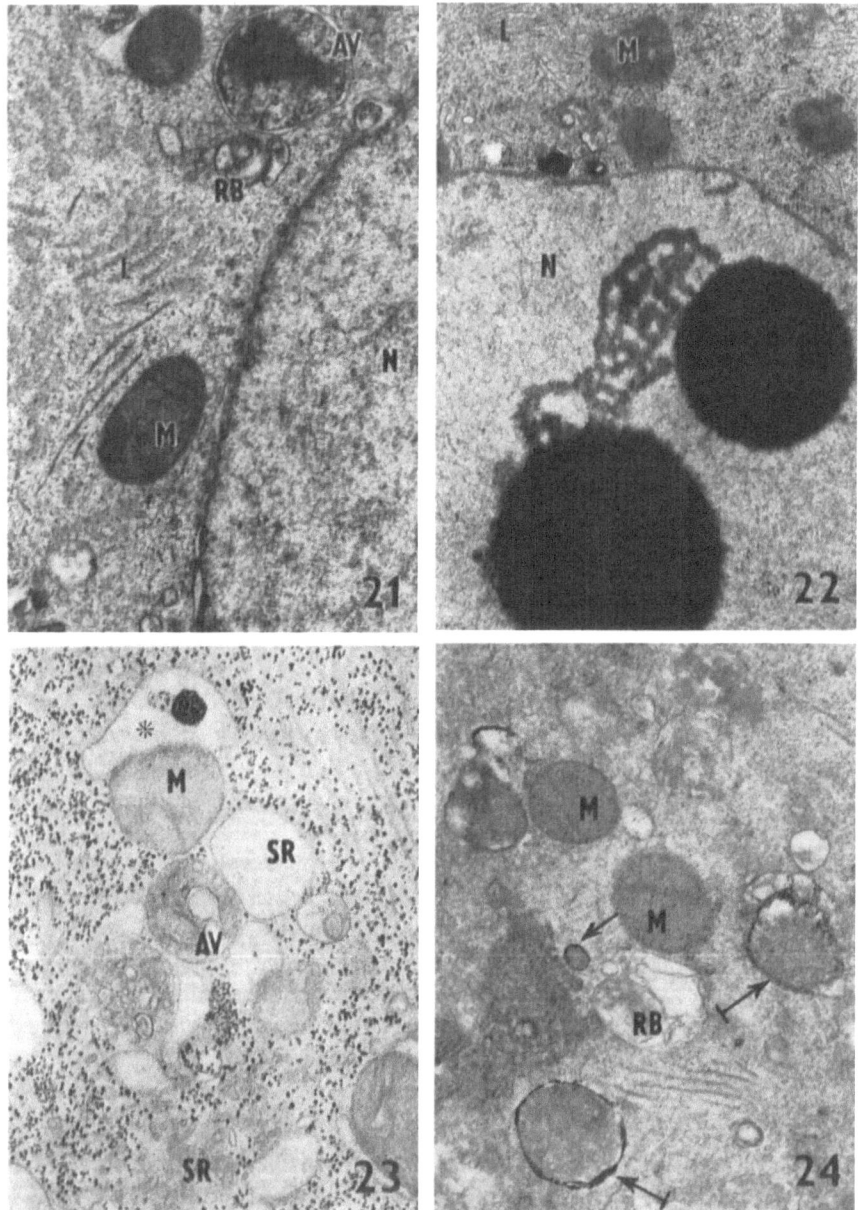

Figs. 21–24

Fig. 21. Four-cell rat ovum. Part of blastomere with nucleus (N) and cytoplasm containing
mitochondria (M), autophagic vacuole (AV), residual body (RB), and lamellar structures (L).
Fixation glutaraldehyde and OsO$_4$; embedding medium Durcupan ACM; magnification X 22,000

Fig. 22. Four-cell rat ovum. Part of a blastomere with reticulated nucleoli in the nucleus (N) and
mitochondria (M) and lamellar structures (L) in the cytoplasm. Fixation glutaraldehyde and OsO$_4$;
embedding medium Durcupan ACM; magnification X 12,000. From Dvořák, M., Scripta med. 47,
497–502 (1974)

rial, but the bodies can also be empty. Dark multivesicular bodies contain a great number of vesicle-like structures which are light, while dark matrix forms the background. There is always a membrane on the surface of multivesicular bodies. Multivesicular bodies occur in the vicinity of other cell organelles, particularly of vesicles of smooth endoplasmic reticulum and, at this stage, also of secondary lysosomes. Only in individual multivesicular bodies of this stage has acid phosphatase (Šťastná 1977a) and alkaline phosphatase activity (Šťastná, 1979) been proved.

2.3.2.3. Autophagic Vacuoles

Autophagic vacuoles comprise 1.8% of the cytoplasmic volume (Dvořák et al., 1977). They are structures of different sizes, often larger than mitochondria and can even measure more than 1 μm. They are generally formed by membrane structures at different degrees of disintegration (Fig. 23). In some cases, their content resembles that of multivesicular bodies but differs by containing structures other than vesicular ones. They occur both near the nucleus and on the periphery of the cytoplasm. In autophagic vacuoles, organophosphate-resistant esterase activity (Trávník, 1977b, 1978) and sometimes acid phosphatase activity (Fig. 24) has been proved (Šťastná, 1977a), as well as alkaline phosphatase activity (Šťastná, 1979).

2.3.2.4. Residual Bodies

This group includes structures consisting of intensely electron dense material in which none of the characteristic cytoplasmic structures can be distinguished. In these bodies, structures displaying the form of pseudomyelin figures are often present. Residual bodies are smaller than 0.3 μm. They only rarely showed acid phosphatase (Šťastná, 1977a) and alkaline phosphatase activity (Šťastná, 1979).

2.3.2.5. Smooth Endoplasmic Reticulum

Smooth endoplasmic reticulum forms 1.3% of the cytoplasmic volume (Dvořák et al., 1977). It is present near where other cell organelles accumulate. Its components are chiefly vesicles several tenths of microns in size, containing medium electron dense material. Small vesicles and/or short tubules are present in a low number and there are always only a few together (Figs. 21 and 24); thus, they do not form large fields, as was the case in the preceding stages. In vesicles of smooth endoplasmic reticulum, organophosphate-sensitive and partly also organophosphate-resistant esterases (Trávník 1977b, 1978) have been proved. Acid and alkaline phosphatase activity have been found in smooth endoplasmic reticulum in only a relatively small part of four-cell stages (Šťastná, 1977a, 1979).

Fig. 23. Four-cell rat ovum. Demonstration of glycogen particles. Mitochondria (M); smooth endoplasmic reticulum (SR); smooth sac with glycogen (*); autophagic vacuole (A V). Fixation glutaraldehyde and OsO₄; embedding medium Durcupan ACM; nonstained ultrathin section; magnification X 26,000

Fig. 24. Four-cell rat ovum. Demonstration of acid phosphatase activity in primary lysosome (→) and in secondary lysosomes (→). Mitochondria (M); residual body (RB). Fixation glutaraldehyde and OsO₄; embedding medium Durcupan ACM; nonstained ultrathin section; magnification X 26,000

2.3.2.6. Granular Endoplasmic Reticulum

Granular endoplasmic reticulum has not been observed with certainty in the four-cell stage.

2.3.2.7. Ribosomes

Ribosomes are present in a small number, especially in the vicinity of the nucleus and mitochondria. They are arranged in polysomes. Schlafke and Enders (1967) state that a few small clusters of ribosomes are now apparent for the first time.

2.3.2.8. Golgi Complex

The Golgi complex takes up 0.2% of the cytoplasmic volume (Dvořák et al., 1977). Its structures within one Golgi field are not numerous. They are formed by only a few narrow cisternae with individual small vesicles at their ends. Schlafke and Enders (1967) describe it as zones consisting of stacks of parallel lamellae, vesicles, and often one or two dense granules. The Golgi complex is located both on the periphery of the blastomere cytoplasm and in its center. In some four-cell ova, acid phosphatase (Šťastná, 1977a, 1978a) and alkaline phosphatase activity (Šťastná, 1979) has been found in the structures of the Golgi complex.

2.3.2.9. Centrioles

Centrioles have not been found in the material examined.

2.3.2.10. Lamellar Structures

Lamellar structures are a significant component of the cytoplasm, taking up 28.5% of its volume (Dvořák et al., 1977). They are concentrated in large areas in which only individual glycogen particles are present. Lamellae run in parallel bundles in various directions and are often undulated (Fig. 25). They are both near the nucleus and on the periphery of the cytoplasm, but they do not reach the cell membrane. Close to lamellar structures and/or directly on them, weak organophosphate-sensitive esterase activity appears (Trávník, 1977b, 1978).

2.3.2.11. Glycogen

Glycogen is again represented by the monoparticular form whose particles are scattered irregularly in the cytoplasm (Fig. 23). In the region of accumulation of lamellar structures and cell organelles, they occur in a substantially lower number than in the remaining ground cytoplasm. Aggregations of glycogen particles inside smooth vesicles are present, even though in a much lower number. In some cases, it was possible to observe that the glycogen complex located in the center of a smooth-surfaced vesicle was characterized by a rather amorphous structure, so that the distinction of the individual glycogen particles was difficult. In addition, very frequently the total electron density of this complex was reduced and very rarely even smooth vesicles were found containing only islets of ground cytoplasm without an aggregation of glycogen particles (Čech, 1977a). In multivesicular and residual bodies, glycogen was only sporadically found. Aggregations of glycogen particles, mostly enveloped by a membrane, were observed in single cases even outside the blastomeres, below the zona pellucida (Čech, 1977a).

2.3.2.12. Lipid Droplets

Lipid droplets have not been found in any of the four-cell ova examined. This circumstance does not, however, exclude their occasional occurrence.

2.3.2.13. Cell Membrane

The cell membrane, mostly on the free surfaces of the ovum, projects into short microvilli. The microvilli are of different lengths up to 0.3 μm and are covered, as well as the other part of the cell membrane, by the glycocalyx. The intercellular space between the blastomeres is, for the most part, only several tens of nanometers wide (Fig. 25). The cell membrane is mostly smooth there. Microvilli occur only in the places where the intercellular space is broader. It is constantly wider where more (three) blastomeres are in contact.

Differentiated junctional structures are not yet present at this stage. The first sign of the formation of structures increasing adhesion can be considered the occurrence of spots of dense material near the inner surface of the cell membrane (Fig. 25), situated more often on one side than on both sides with respect to the intercellular space (Dvořák, 1977). The activity of alkaline phosphatase (Šťastná, 1979) has been found on the cell membrane of some four-cell ova.

2.3.2.14. Cortical Granules

Cortical granules found in only one case in the four-cell stage had a typical structure and location below the cell membrane. They were not present in the other ova of this stage.

2.3.3. Zona Pellucida

The zona pellucida is about 2 μm wide and stains well with ruthenium red. It is separated from the cell membrane by the perivitelline space, 0.15 μm wide, into which microvilli project from the surface of the ovum. Szollosi (1975a) observed "periodisch strukturierte Körper" consisting probably of acid glycoproteins in the perivitelline space.

2.4. Eight-Cell Ovum

The eight-cell ova (Fig. 26) were obtained at 7:00 a.m. - 9:00 a.m. on the 4th day of pregnancy. At this stage, the ovum has mostly the shape of a flattened disc whose thickness is one to two layers of blastomeres. But even such a space arrangement as described by Schlafke and Enders (1967) is no exception. These authors state that the cells are more commonly arranged in a triradiate fashion, two or three cells long and only two cells wide in any direction. The blasomeres are of polyhedral shape and approximately the same size, their diameters being 31 μm. On the other hand, Schlafke and Enders (1967) claim that considerable variation in size of individual blasto-

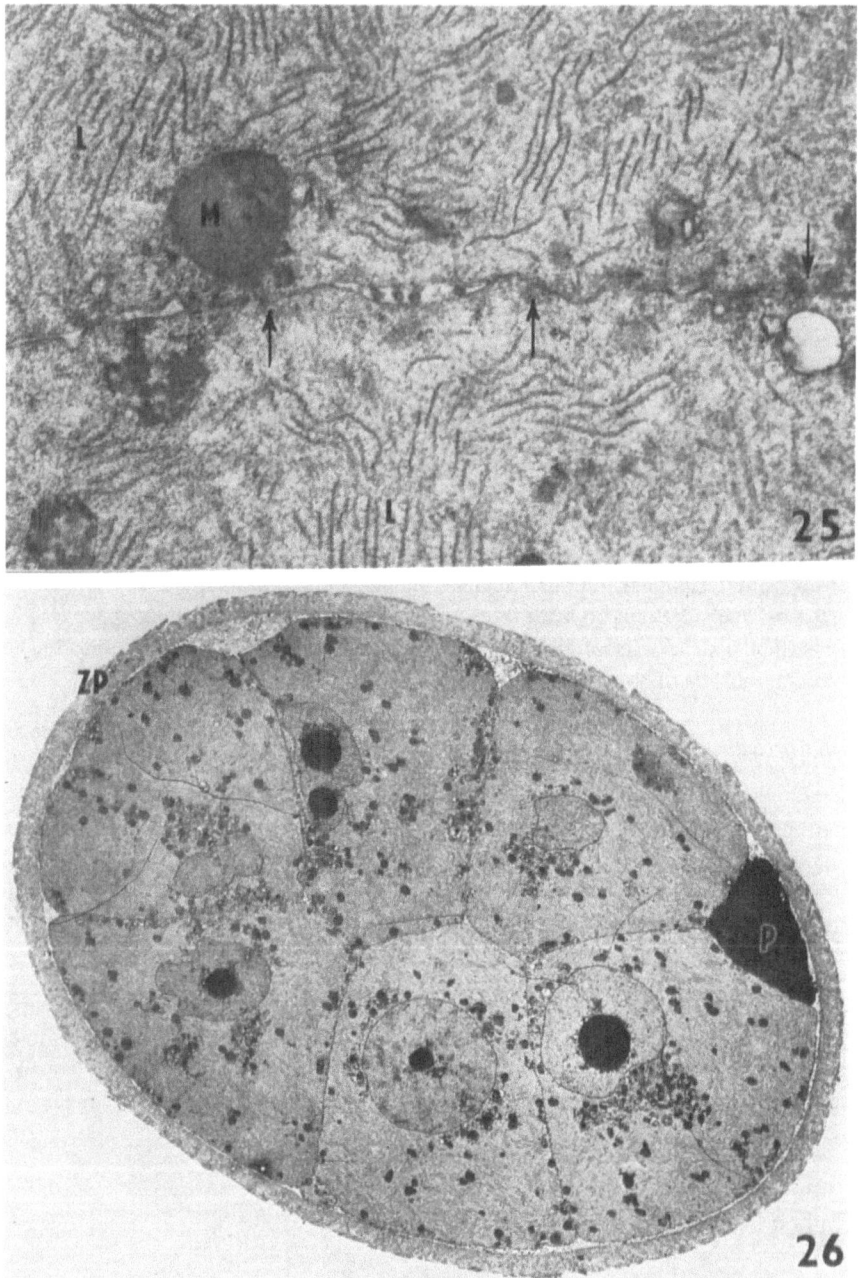

Fig. 25. Four-cell rat ovum. Contact of two blastomeres with small condensations of dense material at the cell membrane (→); mitochondrion (M); lamellar structures (L). Fixation glutaraldehyde and OsO₄; embedding medium Durcupan ACM; magnification X 28,000

Fig. 26. Eight-cell rat ovum. In the section, six blastomeres and one polar body (P) are visible; zona pellucida (ZP). Fixation glutaraldehyde and OsO₄; embedding medium Durcupan ACM; magnification X 1,700

meres occurs, but in no instance could two distinctly different groups of cells be sorted out on a size basis. Blastomeres are mostly closely attached to each other, the perivitelline space being narrow and only in places mildly broadened. Those extensions can, however, be artificial, due to the shrinkage of the blastomeres during fixation. There is an intact zona pellucida on the surface of the eight-cell ovum. The size of the eight-cell ovum is approximately 65 μm in length and 45 μm in width. Each blastomere consists of the nucleus and cytoplasm.

2.4.1. Nucleus

The nuclei are round or oval, their size being on the average 9.8 μm. Their surfaces are either smooth or small invaginations can be observed. The nuclei are located mostly excentrically in the blastomeres and consist of a nuclear envelope, chromatin, and nucleoli. Intranuclear annulate lamellae are sometimes present.

2.4.1.1. Nuclear Envelope

The nuclear envelope has the usual structure being formed by two membranes between which there is the perinuclear space with a regular width of 15 - 20 nm (Fig. 27). In some blastomeres, irregularities in the width of the nuclear envelope were observed which were due to the fact that signs of the blebbing process were clearly visible on the outer membrane. Intranuclear annulate lamellae with pores filled with amorphous electron dense fibrillar material spread from the inner membrane into the interior of the nucleus to different distances (Fig. 27). The nuclear envelope is provided with numerous pores filled with electron dense amorphous material. On the outer membrane of the nuclear envelope, a small quantity of ribosomes are attached, which is in accordance with the data by Schlafke and Enders (1967). In the perinuclear space at this stage, organophosphate-sensitive esterase activity (Trávník, 1977b, 1978) was still found; in some ova (Fig. 28) acid phosphatase (Šťastná, 1977a) and alkaline phosphatase activity (Šťastná, 1979) was also observed.

2.4.1.2. Chromatin

Nuclear chromatin is distributed evenly in the nuclei of the blastomeres, causing the nuclei to appear light.

2.4.1.3. Nucleolus

In ultrathin sections, most often one or two nucleoli are found in the blastomere nucleus (Figs. 26 and 27), their size being about 3 - 4 μm. They are mostly formed by the central pars fibrosa surrounded by the pars granulosa. As for the extent, the pars fibrosa, as a rule, still prevails over the pars granulosa. Nucleoli of the compact type (2.5%) were only observed individually; on the other hand, the number of reticular nucleoli was also low (10%) (Dvořák, 1974a). Small aggregations of material of similar appearance and density as that of the pars granulosa are often located isolated, independent of the nucleolus, in the karyoplasm and often attached to the nuclear envelope. Schuchner (1970) states that at this stage the periphery of the nucleolus changes. The nucleolus acquires an uneven surface, the prominent fields having a fibrillar and granular structure.

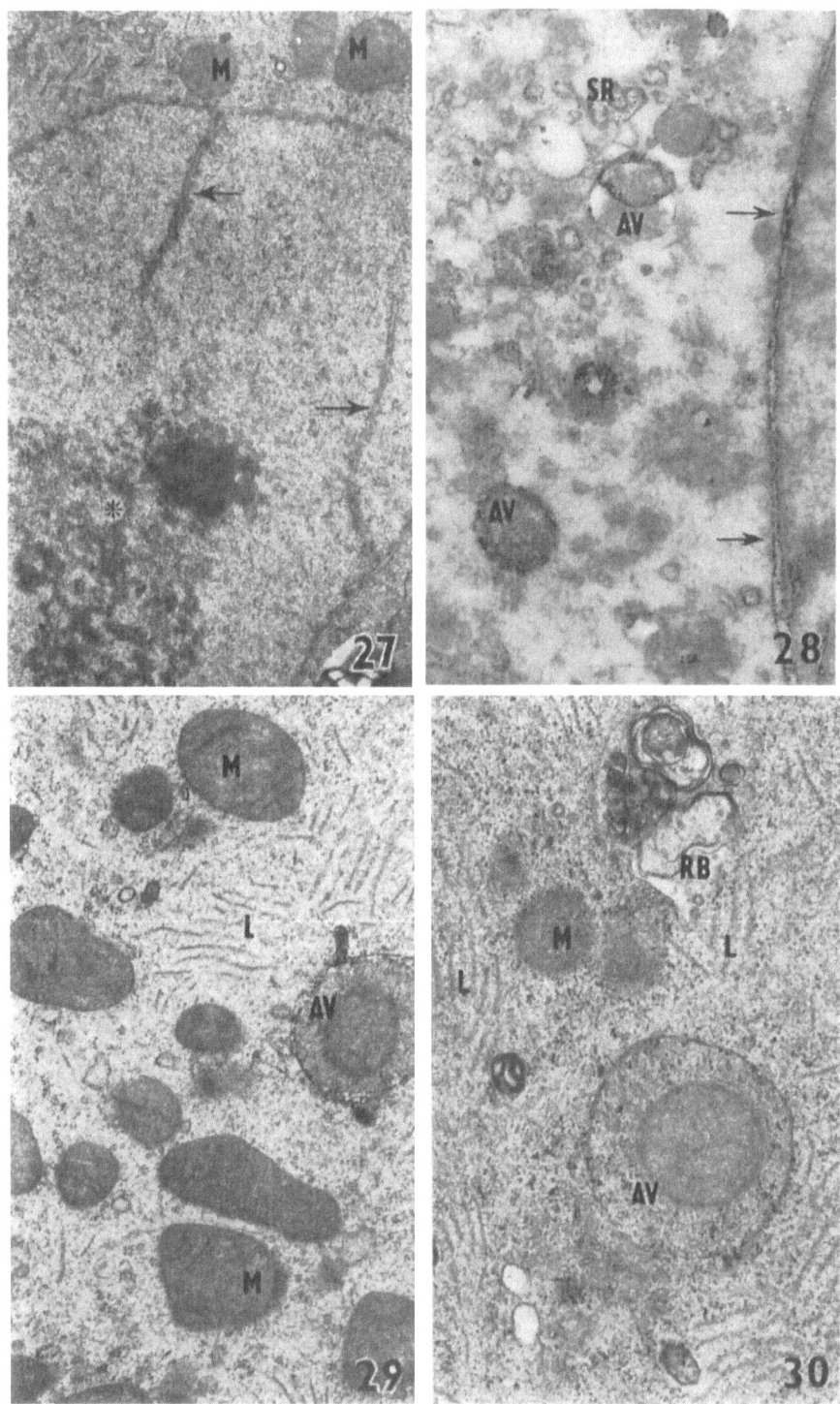

Figs. 27–30

2.4.2. Cytoplasm

Cell organelles located in the ground cytoplasm are not distributed regularly. Most of them are located in an extensive region near one pole of the nucleus (Fig. 26). In the region of accumulation of cell organelles, lamellar structures are only present in small quantities in the ground cytoplasm. Some cell organelles, particularly mitochondria, are often distributed in the cytoplasm individually or in groups. Large regions of ground cytoplasm are, however, filled with lamellar structures, and cell organelles are missing. The distribution pattern of cell organelles precedes the actual segregation, as is encountered in later stages of development.

2.4.2.1. Mitochondria

Mitochondria (Figs. 29 and 30) take up 5.2% of the cytoplasmic volume (Dvořák et al., 1977). They are principally located in the juxtanuclear aggregation of organelles but also individually or in groups elsewhere in the cytoplasm. They are usually round to oval, with a diameter of about 0.5 - 0.7 μm, and contain dense matrix and comparatively few mitochondrial cristae, often arch-like oriented and joined on both ends to the inner mitochondrial membrane. In mitochondria, cristae are often formed by three membranes, the middle one being thicker than the outer two. These specially formed cristae can be evidence of new formation or fusion of mitochondrial cristae. Schlafke and Enders (1967) mention such cristae as "fused cristae."

By their morphologic features, the mitochondria of this stage completely resemble the mitochondria of the earlier stages. Some longer mitochondria are only present individually in the blastomeres of this stage whose width is smaller (ca. 0.2 μm). Several parallel cristae are usually located in them, oriented perpendicular to the longitudinal axis of the mitochondrion, as a rule near one pole. Schlafke and Enders (1967) state that many of the mitochondria are now elongated with lamellar cristae normal for their long axis.

The most important difference from the preceding stages is the fact that most mitochondria of this stage, and particularly those located in the vicinity of the nucleus, have regularly close topographic relations to granular endoplasmic reticulum. There are usually one or several short cisternae of granular endoplasmic reticulum which are at-

---

Fig. 27. Eight-cell rat ovum. Part of a blastomere with reticulated nucleolus (∗) and intranuclear annulate lamellae (→) in the nucleus; in the cytoplasm, mitochondria (M) and lamellar structures are present. Fixation glutaraldehyde and OsO_4; embedding medium Durcupan ACM; magnification X 20,000

Fig. 28. Eight-cell rat ovum. Demonstration of alkaline phosphatase activity in the perinuclear space (→), smooth endoplasmic reticulum (SR) and autophagic vacuole (AV). Fixation glutaraldehyde and OsO_4; embedding medium Durcupan ACM; nonstained ultrathin section; magnification X 41,000

Fig. 29. Eight-cell rat ovum. Cytoplasm with mitochondria (M), autophagic vacuole (AV), and lamellar structures (L). Fixation glutaraldehyde and OsO_4; embedding medium Durcupan ACM; magnification X 21,000

Fig. 30. Eight-cell rat ovum. Cytoplasm with autophagic vacuole (AV), residual body (RB), mitochondrion (M) and lamellar structures (L). Fixation glutaraldehyde and OsO_4; embedding medium Durcupan ACM; magnification X 40,000

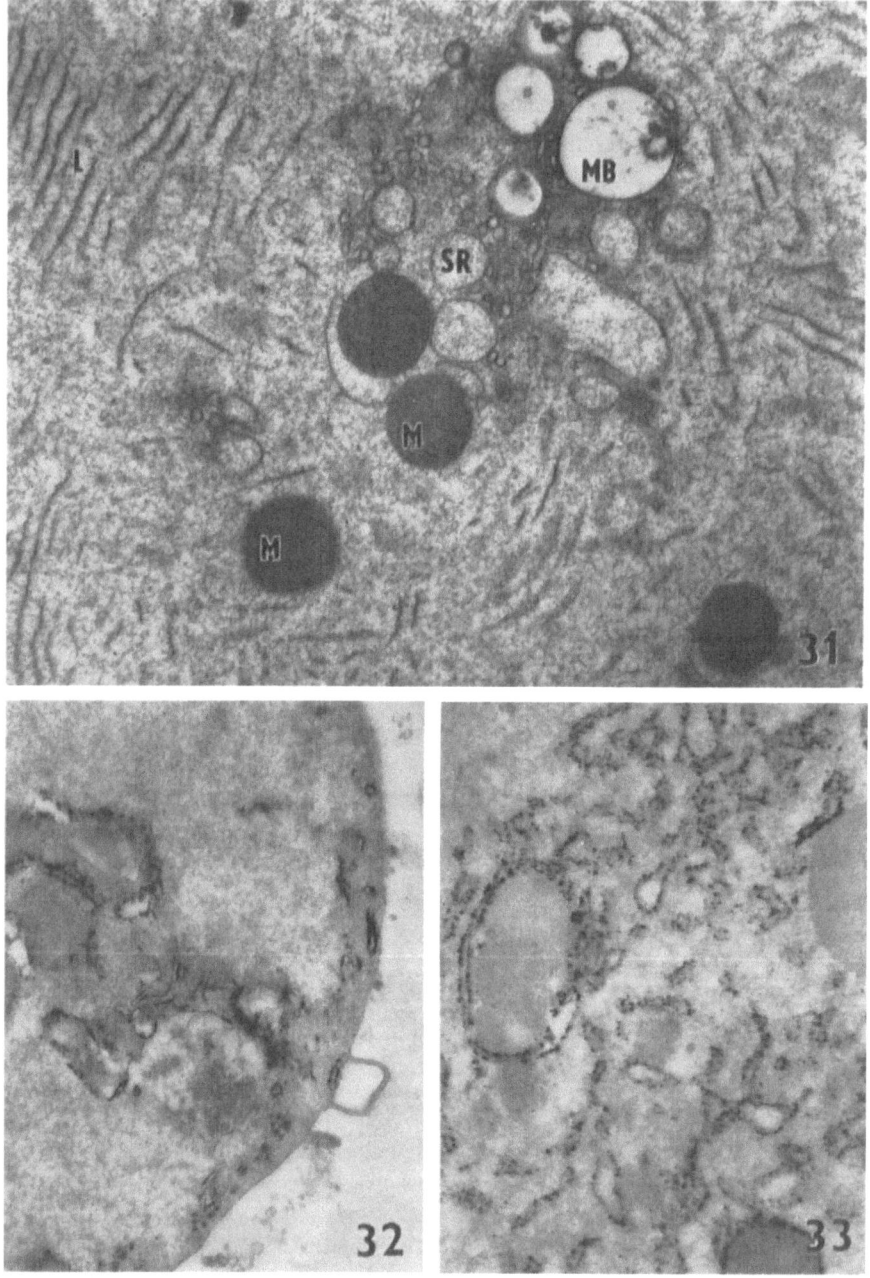

Figs. 31–33

Fig. 31. Eight-cell rat ovum. Part of the cytoplasm with multivesicular bodies (*MB*), mitochondria (*M*), smooth endoplasmic reticulum (*SR*), and lamellar structures (*L*). Fixation glutaraldehyde and OsO$_4$; embedding medium Durcupan ACM; magnification X 23,000

Fig. 32. Eight-cell rat ovum. Demonstration of the organophosphate-sensitive nonspecific esterase in the smooth vesicles. Fixation formaldehyde and OsO$_4$; embedding medium Durcupan ACM; nonstained ultrathin section; magnification X 26,000

tached to these mitochondria and which lack ribosomes on the side facing the mitochondria. Smooth sacs or tubules are still attached to some mitochondria (Fig. 31), as
also stated by Schlafke and Enders (1967). In the mitochondria of this stage, succinate
dehydrogenase activity (Šťastná, 1977b, 1978b) has been proved, in some cases even
organophosphate-sensitive esterase activity (Trávník, 1977b, 1978).

2.4.2.2. Multivesicular Bodies

In comparison with the earlier stages, the number of multivesicular bodies is substantially reduced (Fig. 31), and in this stage it accounts for only 0.43% of the cytoplasmic
volume of the blastomere (Dvořák et al., 1977). Their morphologic picture and the
results of acid phosphatase examination are the same as in the earlier stages.

2.4.2.3. Autophagic Vacuoles

Autophagic vacuoles are very numerous in this stage (Figs. 29 and 30). They are limited by one or two membranes and their size reaches $1 - 2 \mu m$. They contain ground cytoplasm in which various cell structures, such as mitochondria, endoplasmic reticulum,
the Golgi structures, ribosomes, glycogen, and lamellar structures are located at different degrees of degradation. Frequent origin of autophagic vacuoles at this stage is
evidenced by the fact that they often contain nearly hitherto intact mitochondria or
other organelles. The proof of nonspecific acid phosphatase (Šťastná, 1977a) and
organophosphate-resistant esterase activity (Trávník, 1977b, 1978) in some autophagic
vacuoles supports their ranking among secondary lysosomes.

2.4.2.4. Residual Bodies

Residual bodies (Fig. 30) are found in the cytoplasm of the blastomeres in a high
number, and, together with autophagic vacuoles, they comprise 4.1% of the cytoplasmic volume (Dvořák et al., 1977), i.e., almost the same volume as mitochondria. Residual bodies are mostly located in an aggregation of cell organelles near one pole of the
nucleus. They reach a size of $1 - 2 \mu m$ and are limited by a single membrane. A high
number of membrane fragments and myelin figures, vesicles, vacuoles, and granules of
different sizes, and a various amount of dense amorphous material are typical. Acid
phosphatase activity (Šťastná, 1977a) has been found in them.

2.4.2.5. Smooth Endoplasmic Reticulum

The amount of structures of smooth endoplasmic reticulum has dropped in comparison
with the earlier stages and only forms 0.41% of the cytoplasmic volume. In particular,
complexes of vesicles and tubules of smooth endoplasmic reticulum, typical of earlier
stages, have disappeared from the blastomeres. Only small groups of smooth vesicles
located in the vicinity of the other cell organelles and/or single vesicles and tubules
located in the cytoplasm which are either free or attached to mitochondria have remained from these formations. In some of the vesicles (Figs. 32 and 33), organophos-

---

Fig. 33. Eight-cell rat ovum. Demonstration of alkaline phosphatase in the cisternae of smooth
endoplasmic reticulum. Fixation glutaraldehyde and OsO_4; embedding medium Durcupan ACM;
nonstained ultrathin section; magnification X 25,000

phate-sensitive esterase activity (Trávník, 1977b, 1978) has been proved, as well as that of acid phosphatase (Šťastná, 1977a) and alkaline phosphatase (Šťastná, 1979).

2.4.2.6. Granular Endoplasmic Reticulum

The presence of granular endoplasmic reticulum was first definitely found at this stage. Its structures take up 0.4% of the cytoplasmic volume (Dvořák et al., 1977) and are chiefly located in the perinuclear region and the zone containing organelles (Figs. 29 and 30). They can also be found near groups of mitochondria and occasionally elsewhere in the cytoplasm. Granular endoplasmic reticulum has the form of short tubules or vesicles with a few ribosomes distributed at irregular distances on the membranes. Schlafke and Enders (1967) state that granular endoplasmic reticulum has a narrower space between the membranes than that seen in previous stages. Part of its structures are, as stated above, attached closely to mitochondria. In some tubules and vesicles of granular endoplasmic reticulum, organophosphate-sensitive esterase activity (Trávník, 1977b, 1978) has been found, as well as alkaline phosphatase activity (Šťastná, 1979).

2.4.2.7. Ribosomes

In comparison with the previous stages, ribosomes evidently increase in number. They are located in the perinuclear region, in the zone containing organelles and, as a rule, close to mitochondria. They have the form of polysomes. Schlafke and Enders (1967) state that polysomes are present more frequently then individual ribosomes. In zones where free ribosomes are located, advanced degradation of lamellar structures is clearly visible. Remnants of lamellar structures are only sporadically preserved. The ground cytoplasm there is filled with medium electron dense filamentous and fine granular material which can come from the degraded lamellae. Some of the ribosomes are, as stated before, bound to structures of granular endoplasmic reticulum and the outer nuclear membrane.

2.4.2.8. Golgi Complex

The Golgi complex is located in areas of organelle aggregations without a conspicuous relation to the blastomere nucleus. It is formed by individual dictyosomes or groups of dictyosomes whose structure is similar to those in the previous stages. The individual dictyosomes consist of three to five short cisternae running straight or curved. A variable number of small vesicles of heterogenous appearance are located near both ends of the cisternae and on the outer and inner surfaces of the Golgi complex. At the outer surface of the Golgi complex, a flat cisterna of granular endoplasmic reticulum is located, studded with ribosomes only on the surface not facing the dictyosome. In the vicinity of the dictyosomes, secondary lysosomes are often present, particularly autophagic vacuoles and residual bodies and vesicle groups of smooth endoplasmic reticulum. In some specimens of the eight-cell stage, acid and alkaline phosphatase activity (Šťastná, 1977a, 1978a) has been proved in the structures of the Golgi complex.

2.4.2.9. Centrioles

Centrioles have not been observed in the blastomeres of the eight-cell stage.

2.4.2.10. Lamellar Structures

Lamellar structures (Figs. 31 and 34) occupy a considerable part of the cytoplasm, taking up 20% of its volume (Dvořák et al., 1977). In the zones containing cell organelles, especially in regions where polysomes appear in the ground cytoplasm, lamellar structures are either altogether missing or only their fragments are present. In the other regions, they have the same arrangement and submicroscopic structure as in the earlier stages.

2.4.2.11. Glycogen

The amount of glycogen particles in the cytoplasm of the blastomeres of the eight-cell stage decreased significantly, and local differences in the frequency of their occurrence are by no means conspicuous. Characteristic is the loss of size uniformity of glycogen particles which very often aggregate into small clusters. Smooth vesicles with glycogen are already missing. One of the typical signs of eight-cell ova is the accumulation of glycogen particles inside autophagic vacuoles, mainly in their peripheral zones (Fig. 35). Multivesicular bodies only rarely contain glycogen particles, while in residual bodies they occur in the same frequency as in the previous stages. Extracellular aggregations of glycogen are relatively frequent and are found mainly between the blastomeres and the zona pellucida. The decrease in the amount of glycogen and its distribution pertains to all blastomeres of the eight-cell stage; individual differences were not observed in this respect.

2.4.2.12. Lipid Droplets

Lipid droplets are regularly present in the blastomeres of the eight-cell stage. They are mostly located individually in the zone containing organelles or near the surface of the blastomeres. They comprise 0.2% of the volume of blastomere cytoplasm (Dvořák et al., 1977).

2.4.2.13. Cell Membrane

The blastomeres of the eight-cell stage are mostly closely attached to each other. Their surface is either quite smooth and the intercellular space very narrow or in some cases, irregular interdigitations which enable the cells to fit close to each other can be observed. The intercellular space can sometimes be dilated in certain regions, into which the neighboring cells send irregular microvilli. The glycocalyx covering the free surfaces of cells also continues to the lateral surfaces of blastomeres (Fig. 36). It forms a continuous rim of different thickness. The cell membrane projects into microvilli whose distribution on the surface is essentially regular; their lengths, like in the preceding cases, are not even. Cell contacts show an arrangement similar to the preceding stage. Only the intercellular space need not be widened at the place of contact of more blastomeres.

The cell junction is strengthened in several ways. At a different degree of differentiation, zonulae adherentes are present. In the peripheral part of cells in some cases, the cell membranes closely approach each other at "points," or they even appear to be coalesced. The first type of junction is a close junction, which in isolated cases is even extensive; the second type of junction corresponds to focal tight junction. In the widened places of the intercellular space, exceptionally long cytoplasmic processes are

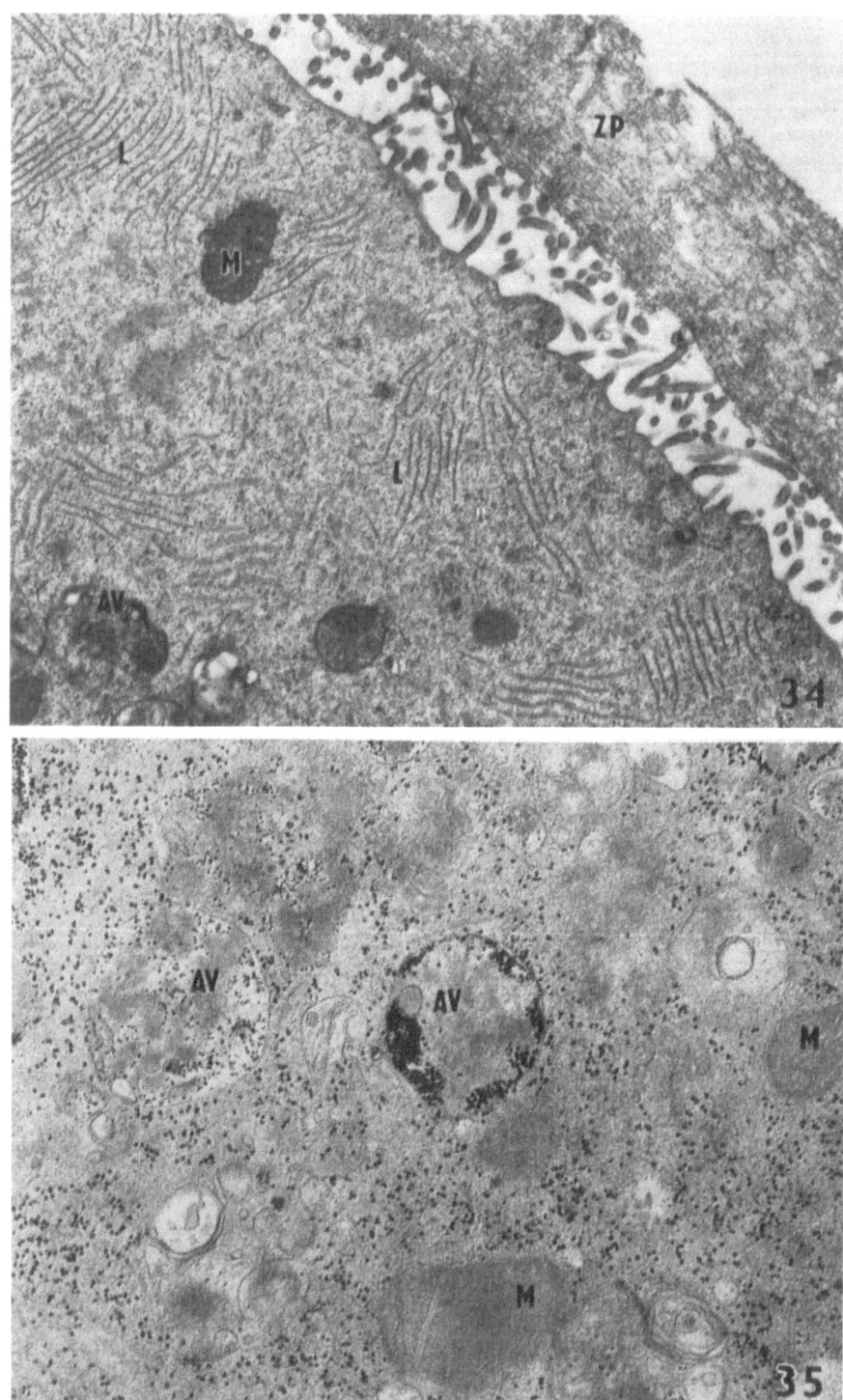

Figs. 34 and 35.

present, leaving one cell and approaching at different distances the cell membrane of the neighboring cell. On the cell membranes of all blastomeres, the activity of nonspecific alkaline phosphatase (Šťastná, 1979) has been found.

2.4.3. Zona Pellucida

The zona pellucida in this ovum shows lower staining affinity to ruthenium red. Deposits in the perivitelline space of the material stained with ruthenium red are very rare.

2.5. Morula

In the rat, the morula (Fig. 37) corresponds to the 16-24-cell stage which is present in the tuba uterina for a short period of time, between 10:00 p.m. - 12:00 p.m. on the 4th day of pregnancy. The early blastocyst develops later. In spite of the fact that some authors do not include the morula as a stage of development in the rat (Schlafke and Enders, 1967), we do so because it is no longer the eight-cell stage and not yet the early blastocyst. Besides, the blastomeres are distinguished at this stage for the first time, some forming the outer layer of cells surrounding the cells forming the inner cell mass. It is necessary to mention that Schlafke and Enders (1967) state that "the basis for blastocyst formation is initiated with the formation of junctional complexes in the eight-cell stage, but the first appearance of evidence of a blastocyst cavity, as apposed to intercellular spaces, is not seen until after another series of cleavages which occurs during the night of day 4."

The morula is elongated and somewhat flattened in shape, and its diameter is (depending on the section plane) 55 - 80 μm. It is composed of polygonal peripheral cells more or less flattened with average diameters of 30 μm. Some peripheral cells are already significantly flat. The cells are closely attached to each other, without signs of the widening of intercellular spaces. This pattern distinguishes the morula from the initial stage of the early blastocyst. The perivitelline space is narrow, separating peripheral blastomeres from the zona pellucida. The mitosis of the blastomeres is asynchronous; in the morulae under investigation, most cells were in the interphase; individual blastomeres were found in mitosis.

2.5.1. Nucleus

The cells always have only one nucleus (Figs. 37 and 38), circular in cross-section or sometimes slightly oval. The nuclear surface is smooth, and it is seldom possible to

Fig. 34. Eight-cell rat ovum. Peripheral part of the cytoplasm with numerous lamellar structures (*L*), autophagic vacuole (*AV*), mitochondrion (*M*), and many microvilli projecting in the perivitelline space; zona pellucida (*ZP*). Fixation glutaraldehyde and OsO$_4$; embedding medium Durcupan ACM; magnification X 20,000

Fig. 35. Eight-cell rat ovum. Demonstration of glycogen particles. Mitochondria (*M*); autophagic vacuole (*AV*). Fixation glutaraldehyde and OsO$_4$; embedding medium Durcupan ACM; nonstained ultrathin section; magnification X 40,000

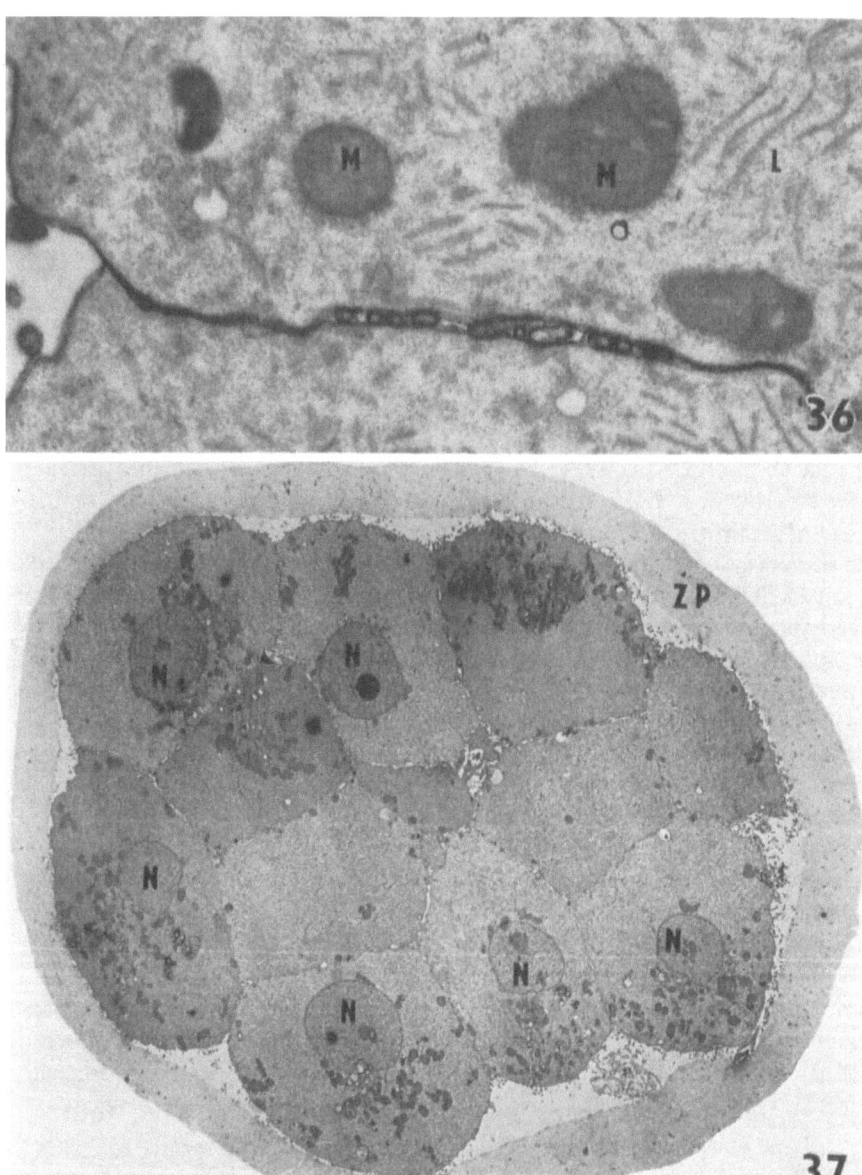

Fig. 36. Eight-cell rat ovum. Contact of peripheral parts of two blastomeres. Glycocalyx is stained dark with ruthenium red; mitochondria (M); lamellar structures (L). Fixation glutaraldehyde and OsO₄; embedding medium Durcupan ACM; nonstained ultrathin section; magnification X 25,000

Fig. 37. Rat morula. Ultrathin section of 12 blastomeres; nuclei (N); zona pellucida (ZP). Fixation glutaraldehyde and OsO₄; embedding medium Durcupan ACM; magnification X 1,800

observe shallow folds of the nuclear envelope. The average diameter of the nuclei is 7 μm. The nuclei are mostly located in the cells centrally, but in some cases also excentrically. In an overall view, they appear light and posses a distinct nucleolus. They consist of a nuclear envelope, chromatin, and nucleoli.

2.5.1.1. Nuclear Envelope

The nuclear envelope (Fig. 38) is formed by two membranes with a narrow perinuclear space, 15 - 20 nm wide. In the nuclear envelope, numerous nuclear pores are present in which — at higher magnification — amorphous electron dense material can be seen. Intranuclear annulate lamellae are derivatives of the inner nuclear membrane, and their mutual connection can be observed even in ultrathin cross-sections. The outer membrane is studded with ribosomes distributed with different density. The total number of ribosomes is higher here than in the eight-cell stage. A direct connection of the outer membrane of the nuclear envelope to cisternae of granular endoplasmic reticulum was sporadically observed. In the vicinity of the nucleus, there is an increased occurrence of polysomes grouped in large aggregations.

2.5.1.2. Chromatin

Nuclear chromatin is distributed evenly in the nucleus and forms small aggregations located regularly about the whole area of the nucleus only sporadically.

2.5.1.3. Nucleolus

A component of the nucleus is mostly one nucleolus, sometimes two. They are large, mostly of reticulated type (Fig. 38). From the original, compact nucleolus, only small spheric formations have been left, around which a nucleolonema is located. The pars granulosa begins to prevail over the pars fibrosa.

2.5.2 Cytoplasm

According to the total morphologic appearance of the cytoplasm, it is possible to characterize some cells as dark, in which the ground cytoplasm reveals a higher electron density than in the other cells. Cells with conspicuously light cytoplasm were always observed in morulae, even though their occurrence was quite rare. Both nuclei and cytoplasmic structures in these cells showed signs of degradation to different degrees. Most of the cells, however, displayed a cytoplasmic structure, as described below. In cytoplasm, the cell organelles are mostly concentrated in the zone around the nucleus and in the peripheral zone near the cell membrane. The central zone of the cytoplasm is occupied by ground cytoplasm with lamellar structures; cell organelles in this zone are present only in places.

2.5.2.1. Mitochondria

Mitochondria amount to 4.9% of the cytoplasmic volume (Dvořák et al., 1977). Most of them are located in zones in which cell organelles occur (Fig. 39); individual mitochondria are, however, also located in regions of occurrence of lamellar structures. The size and shape of mitochondria varies, but the majority of mitochondria are slightly oval (round mitochondria are less frequent than in the earlier stages). The matrix is

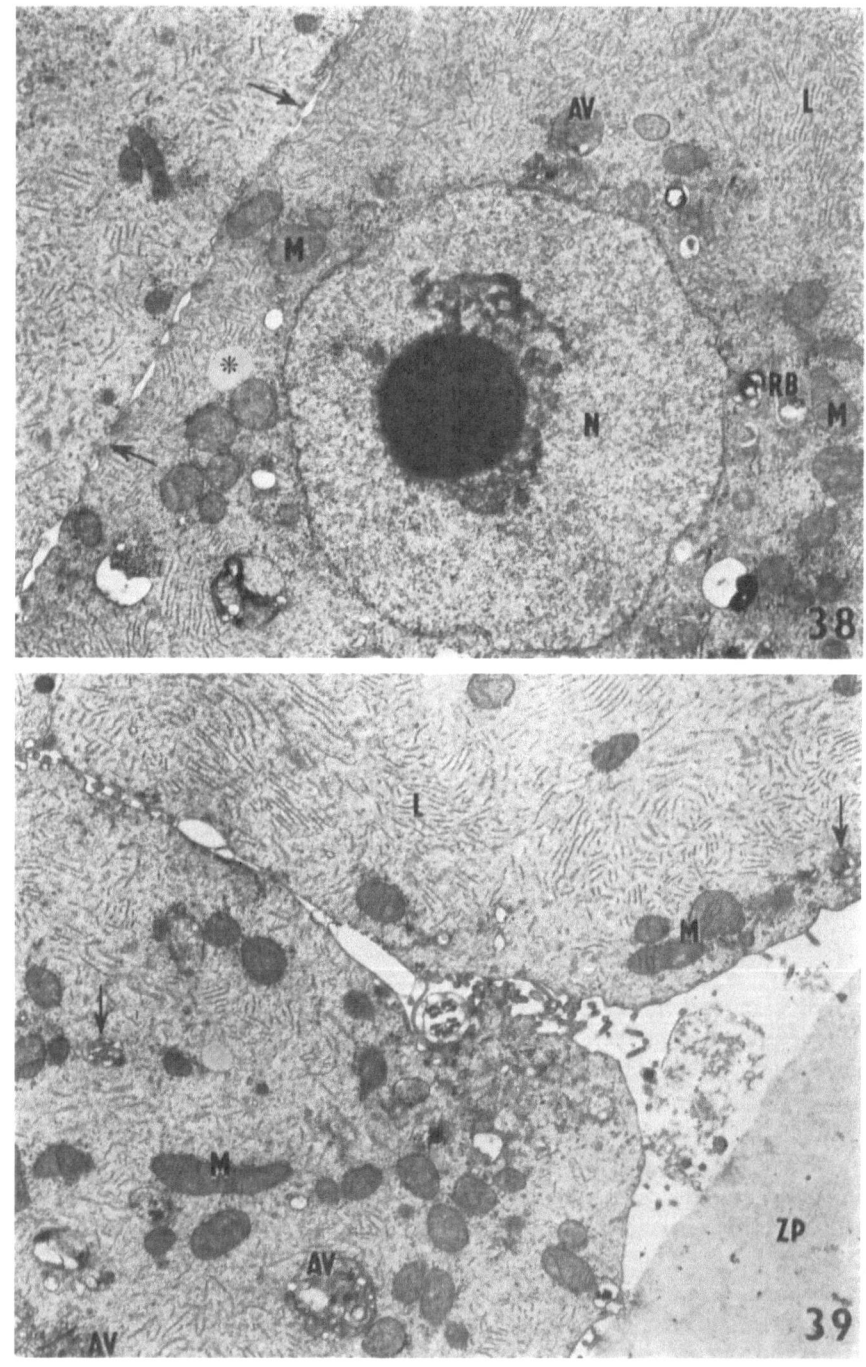

Figs. 38 and 39

Fig. 38. Rat morula. Contact of two blastomeres with narrow intercellular space (→). Nucleus (N) with distinct reticulated nucleolus, mitochondria (M), lamellar structures (L), autophagic vacuole (AV), residual body (RB), and lipid droplet (*). Fixation glutaraldehyde and OsO₄; embedding medium Durcupan ACM; magnification X 11,000

very electron dense. Cristae are arch-like and parallely arranged in only some mito-chondria (Fig. 40). Elongated mitochondria with cristae oriented perpendicular to the longitudinal axis are more frequent than in the preceding stage. Some cristae are slight-ly widened and vesicular in shape.

Cisternae of granular endoplasmic reticulum located close to mitochondria are lacking ribosomes at the side opposite the mitochondria. Numerous polysomes also occur in the vicinity of mitochondria.

2.5.2.2. Multivesicular Bodies

Multivesicular bodies occur at this stage only as an exception. Multivesicular bodies with numerous elementary vesicles usually found in the previous stages were never ob-served. Individual tiny vesicles were observed only rarely. Some bodies whose interior appeared empty were observed.

2.5.2.3. Autophagic Vacuoles

Autophagic vacuoles (Fig. 41) together with residual bodies comprise 3,9% of the cy-toplasmic volume (Dvořák et al., 1977). They are mostly limited by one membrane. Autophagic vacuoles can reach a size of up to 2 μm. Inside them extensive areas of ground cytoplasm with glycogen and lamellar structures are often separated, and a-mong cell organelles, mitochondria of the type usually found in earlier stages, ribo-somes, structures of endoplasmic reticulum, and smooth vesicles of variable sizes are fre-quently observed. The morphologic patterns of many autophagic vacuoles suggest that a great number of them represent the initial forms of the autophagic process. In some autophagic vacuoles, the activity of acid phosphatase (Šťastná, 1977a) has been proved.

2.5.2.4. Residual Bodies

Residual bodies (Fig. 42) are located, the same as in autophagic vacuoles, in the re-gions where cell organelles occur. By their size and a high content of membrane frag-ments and by the positive reaction to acid phosphatase, they resemble the residual bodies of the preceding stage.

2.5.2.5. Smooth Endoplasmic Reticulum

Smooth endoplasmic reticulum occurs rarely and accounts for 0,3% of the cytoplasmic volume (Dvořák et al., 1977). They are small vesicles 0.06 μm in size, individually distributed among cell organelles, or major sacs up to 0.5 μm large, also occurring indi-vidually among lamellar structures.

2.5.2.6. Granular Endoplasmic Reticulum

Granular endoplasmic reticulum forms 0.6% of the cytoplasmic volume (Dvořák et al., 1977). Sacs or vesicles are of various sizes and shapes. Cisternae of granular endoplas-

---

Fig. 39. Rat morula. Peripheral parts of two blastomeres with numerous mitochondria (*M*), auto-phagic vacuoles (*A V*), lamellar structures (*L*), vesicles of smooth endoplasmic reticulum (→); zona pellucida (*ZP*). Fixation glutaraldehyde and OsO$_4$; embedding medium Durcupan ACM; magnifi-cation X 10,000

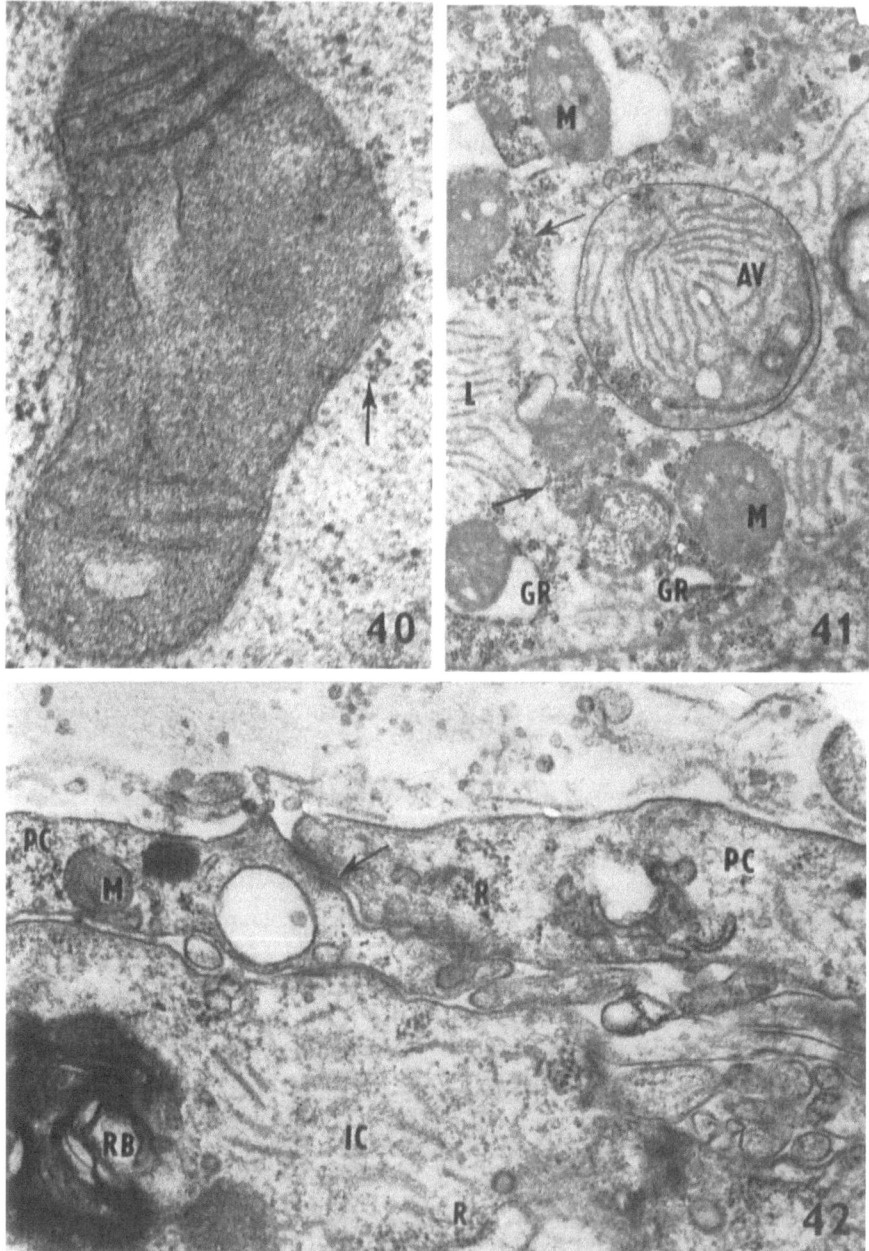

Fig. 40. Rat morula. Mitochondrion with typical arch-like cristae; ribosomes (→). Fixation glutaraldehyde and OsO₄; embedding medium Durcupan ACM; magnification X 76,000

Fig. 41. Rat morula. Part of a cytoplasm with autophagic vacuole (A V), mitochondria (M), ribosomes (→), granular endoplasmic reticulum (GR), lamellar structures (L). Fixation glutaraldehyde and OsO₄; embedding medium Durcupan ACM; magnification X 27,000

Fig. 42. Rat morula. Contact of two peripheral cells (PC) and inner cell (IC). Focal tight junction (→), ribosomes (R), residual body (RB), mitochondrion (M). Fixation glutaraldehyde and OsO₄; embedding medium Durcupan ACM; magnification X 47,000

mic reticulum are flattened or mildly dilated and their cross-sections are longer than in the previous stage. They are situated close to mitochondria and sometimes run from one mitochondrion to the next. Occasionally, branching cisternae have also been observed. Granular endoplasmic reticulum at this stage takes the form of vesicles of various sizes. Accidental direct connection of granular endoplasmic reticulum with the perinuclear space has been observed. The distribution of ribosomes on the membranes of granular endoplasmic reticulum of the morula is considerably irregular.

2.5.2.7. Ribosomes

The number of ribosomes, both free and bound to endoplasmic reticulum, is higher than in the preceding stage. Free ribosomes are arranged in polysomes and distributed near other cell organelles. It is in this stage that they aggregate into clusters for the first time (Fig. 42). They practically never occur in regions with lamellar structures.

2.5.2.8. Golgi Complex

The Golgi complex takes up 0.1% of the cytoplasmic volume (Dvořák et al., 1977). Its structure, location, and relation to other cell orangelles resemble those in the preceding stage. The Golgi complex (field) is mostly small. Often, however, we can see a Golgi complex whose cisternae are significantly longer than in the eight-cell ovum.

2.5.2.9. Centrioles

In the material that was at our disposal, no centrioles were observed, despite the fact that sections were made and studied in incomplete series for morphometric purposes.

2.5.2.10. Lamellar Structures

Lamellar structures occupy large regions of cytoplasm (Figs. 38 and 39) and account for 22.6% of the cytoplasmic volume (Dvořák et al., 1977). In ultrathin sections, they are found either in cross-sections where they appear as short filamentous structures of which several are always located next to each other, or they are cut more or less tangentially where they appear as small plane formations which are not limited and are of unequal density. In contrast to the preceding stages, the lamellar structures are clearly shorter. They are missing in the regions of concentrations of cell organelles and only penetrate on their periphery.

2.5.2.11. Glycogen

The density of glycogen particles in cytoplasm is approximately the same as in the eight-cell ovum. Particles are again characterized by distinct differences in size. Among the lamellar structures, glycogen particles were only rarely present. The occurrence of aggregations of glycogen particles was quite regularly proved inside autophagic vacuoles. Solitary glycogen particles were also found in residual bodies.

2.5.2.12. Lipid Droplets

Lipid particles occur regularly in cytoplasm. They are either individual lipid droplets of greater dimensions – up to 0.6 μm – or small groups of three to five lipid droplets of half the above size.

2.5.2.13. Cell Membrane

The outer surfaces of peripheral cells of the morula are slightly rounded, so that on the surface of the ovum at the site of contact of the two blastomeres deep furrows often occur (Fig. 37). The cell membrane projects into numerous microvilli of different lengths distributed regularly along the whole surface of the ovum. The cell membrane is smooth at the site of cell contacts and only in places, particularly in the region of contact of several blastomeres, projects into microvilli. The cells are closely attached to each other. The connection of cells is strengthened near the surface of peripheral blastomeres by focal tight junction and the zonula adherens (Fig. 42); in other regions of cell contacts of the inner cell mass, there are scattered indications of the zonula adherens.

2.5.3. Zona Pellucida

The ova are surrounded by the zona pellucida with a usual morphologic structure and an average thickness of 3 μm.

2.6. Early Blastocyst

The stage of the blastocyst is remarkable due to the fact that at the time of its formation the blastocyst cavity develops and two basically different kinds of cells are differentiated, an embryoblast and a trophoblast. Once fully developed, the two cells differ from each other in shape, location, fine structure, function, and further developmental prospects. The process of blastocyst differentiation is simultaneously a prerequisite and a part of the preparation for implantation, a new, qualitatively different phase of the differentiation of the embryo and its membranes in the uterus. According to the degree of development of the blastocyst, two types are distinguished: the early blastocyst and the late blastocyst.

Early blastocysts (Fig. 43) were obtained at 7:00 a.m. - 9:00 a.m. on the 5th day of pregnancy. Early blastocysts (60 - 80 μm in size) are surrounded by the intact zona pellucida. They are formed by one layer of flat to cuboidal trophoblast cells surrounding the blastocyst cavity filled with liquid. On the embryonic pole of the blastocyst, an aggregation of polyhedral cells is located, occupying about one-quarter to one-half of its lumen which forms the embryoblast. At this stage, the cells of the embryoblast and the trophoblast differ in their location and shape. No substantial changes in the structure of embryoblast and trophoblast cells have been found under the electron microscope (Mazanec, 1965; Schlafke and Enders, 1967; Šťastná, 1972).

Most of the embryoblast and trophoblast cells were observed during the period of interphase; rarely, both embryoblast and trophoblast cells were observed in mitosis. These observations provide evidence of an asynchronous cell cycle at this stage. Further attention will be paid only to the cells of the embryoblast and the trophoblast during the interphase.

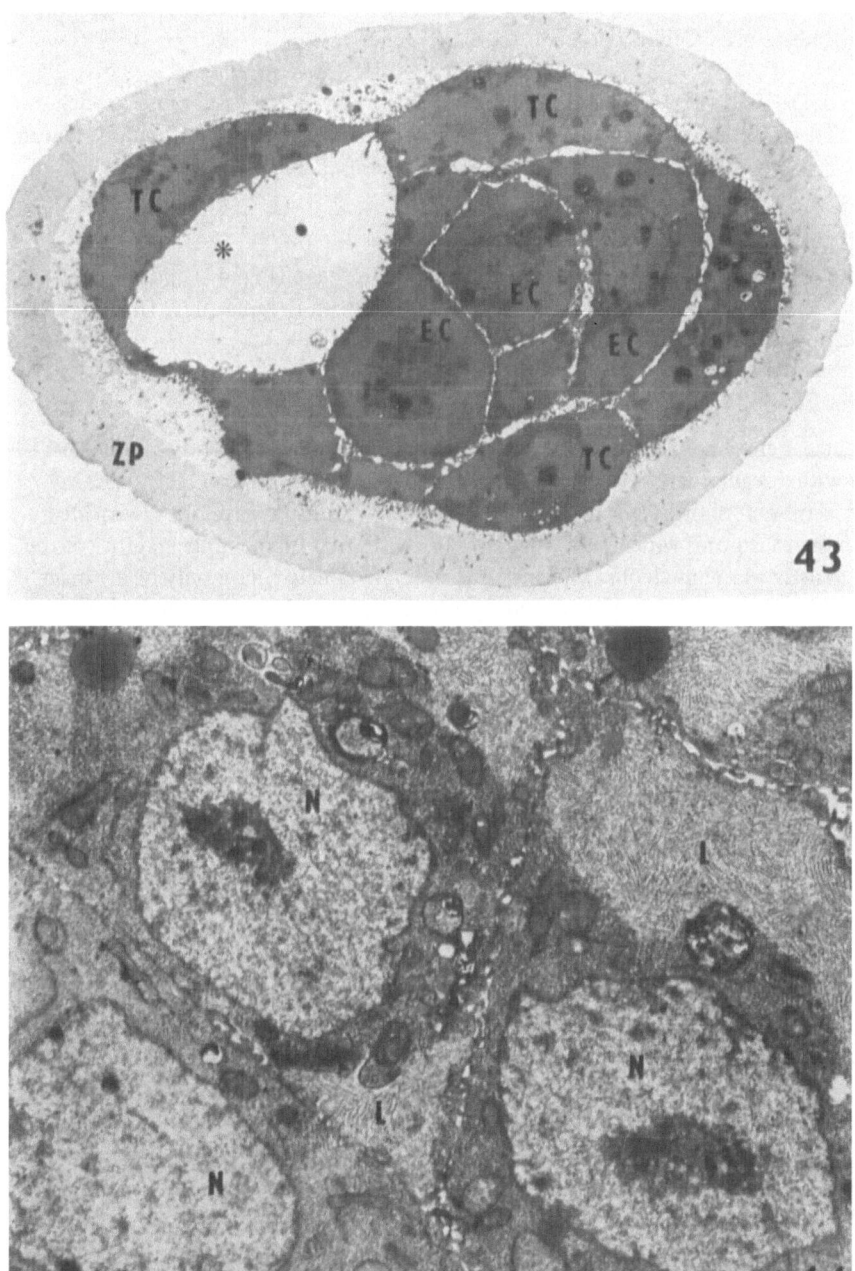

Fig. 43. Early rat blastocyst. Trophoblast cells (*TC*); embryoblast cells (*EC*); blastocyst cavity (∗); zona pellucida (*ZP*). Fixation glutaraldehyde and OsO$_4$; embedding medium Durcupan ACM; magnification X 1800

Fig. 44. Early rat blastocyst. Embryoblast cells with distinct nuclei (*N*) and cytoplasm showing segregation in areas occupied either by lamellar structures (*L*) or cell organelles. Fixation glutaraldehyde and OsO$_4$; embedding medium Durcupan ACM; magnification X 8000

51

2.6.1. Embryoblast Cells

Embryoblast cells (Fig. 44) have a polyhedral shape and are grouped close to each other. They consist of the nucleus and the cytoplasm whose most conspicuous feature is the segregation of cell organelles into certain regions. Their average size is 12.5 μm.

2.6.1.1. Nucleus

The nuclei of embryoblast cells are spheric to oval in shape (Fig. 44) and are 5 μm large. Their surface are either smooth or marked by small irregularities. They are located approximately in the center of embryoblast cells.

Nuclear Envelope

The nuclear envelope is formed by two membranes, between which the perinuclear space with a regular width of 15 - 20 nm is located. In some embryoblast cells, the outer membrane shows signs of blebbing process. The nuclear envelope is provided with numerous pores which, as a rule, are filled with fine filamentous material of the same density as the nucleolus. Intranuclear annulate lamellae have only rarely been observed at this stage, mostly penetrating into the nucleus for only a short distance. They are continuous with the inner membrane of the nuclear envelope and provided with pores. Ribosomes, whose number is higher or on the same level as in the preceding stage, are attached to the outer membrane at irregular distance. Acid phosphatase activity has been proved in the nuclear envelope (Šťastná, 1977a).

Chromatin

Nuclear chromatin is regularly distributed in the nucleus of embryoblast cells in the interphase.

Nucleolus

One to three nucleoli are located in the nucleus of embryoblast cells. The nucleoli (Fig. 45) consist of the pars fibrosa, as a rule forming the central region of the nucleolus, and the pars granulosa, of greater volume than the pars fibrosa, located at the periphery. Small aggregations of material with the same appearance and density as the nucleolus proper are often scattered in the karyoplasm or located near the inner surface of the nuclear envelope.

2.6.1.2. Cytoplasm

In the cytoplasm, the most conspicuous phenomenon is the segregation of organelles (Fig. 44). Cell organelles fill the regions of ground cytoplasm in the vicinity of the nu-

---

Fig. 45. Early rat blastocyst. Contact of two embryoblast cells ($\rightarrow$); nucleus (N); mitochondria (M); Golgi complex (G); ribosomes (R); lamellar structures (L). Fixation glutaraldehyde and OsO_4; embedding medium Durcupan ACM; magnification X 20,000

Fig. 46. Early rat blastocyst. Part of a cytoplasm of embryoblast cell with nucleus (N), mitochondria (M), ribosomes (R), autophagic vacuole (AV), lamellar structures (L), and granular endoplasmic reticulum ($\rightarrow$). Fixation glutaraldehyde and OsO_4; embedding medium Durcupan ACM; magnification X 20,000

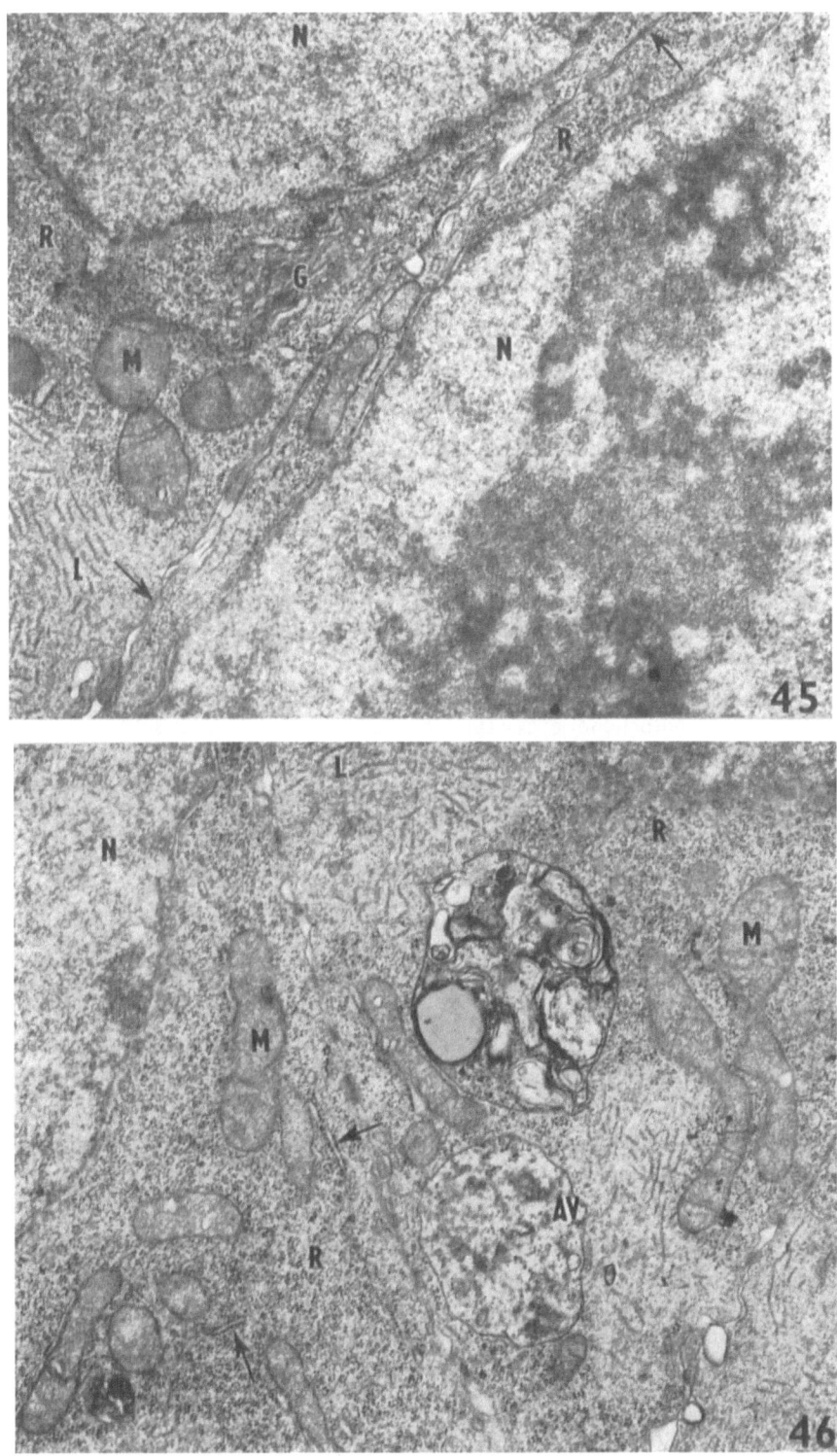

Figs. 45 and 46

53

cleus, on the periphery of embryoblast cells, and in cytoplasmic columns connecting these zones. A considerable part of the cytoplasm is still filled with lamellar structures; in these regions, the cell organelles are missing.

Mitochondria

Mitochondria (Figs. 44 - 46) form 4.7% of the volume of cytoplasm of embryoblast cells (Dvořák et al., 1977). They are located in the regions of the cytoplasm containing organelles. In appearance, size, and shape, certain differences have been observed between the individual blastocysts. Embryoblast cells of some early blastocysts still possess mitochondria similar to those typical of the early stages of cleavage. In most blastocysts, however, mitochondria of oval to rather elongated shape (Fig. 46) are present in the cells of the embryoblast. Often, branched mitochondria constricted in one or several places or even those of bizarre shapes have been observed. Depending on the shape, the length of mitochondria varies from 0.4 - 1.5 μm and thickness from 0.2 - 0.4 μm. The matrix of the mitochondria is medium dense and lighter than in the earlier stages; the number of mitochondrial cristae is higher than in the earlier stages. Most of them are arranged parallely, perpendicular to the longitudinal axis of the mitochondrion, and often form a complete partition of the mitochondrion.

As did Schlafke and Enders (1967), we frequently observed cristae formed by three parallel membranes, the middle one being approximately twice as thick as the other mitochondrial membranes. Mitochondria of the embryoblast cell are topographically closely related to the cisternae of the granular endoplasmic reticulum. These cisternae partly surround the mitochondria; as a rule, they lack ribosomes on the side facing the mitochondria. Occasionally organophosphate-resistant esterase activity (Trávník, 1977b, 1978) has been found in mitochondrial cristae at this stage. In some mitochondria, succinate dehydrogenase activity was present (Šťastná, 1977b).

Autophagic Vacuoles

Autophagic vacuoles (Figs. 44 and 46) are mostly located in zones containing organelles and only rarely in zones occupied by lamellar structures. Two types of autophagic vacuoles can be distinguished: 1) those which principally contain ground cytoplasm with ribosomes and/or other cell organelles (often with considerable numbers of glycogen particles) and 2) those which contain lamellar structures in addition to the ground cytoplasm. Unlike in the earlier stages, mitochondria have not been observed inside autophagic vacuoles. In autophagic vacuoles, acid phosphatase activity (Šťastná, 1977a) and organophosphate-resistant esterase activity (Trávník, 1977b, 1978) has been found.

Residual Bodies

Residual bodies (Fig. 44) together with autophagic vacuoles amount to 3% of the cytoplasmic volume (Dvořák et al., 1977). They are located in the cytoplasm of embryoblast cells in zones containing organelles. They reach a considerable size (average size $\sim$ 1 - 2 μm), are enclosed by a single membrane, and are chiefly characterized by a high content of membrane fragments and myelin figures. In addition, they also contain sacs and vacuoles in variable quantities, granules of different sizes, lipid droplets, and different quantities of amorphous material of various densities difficult to determine.

Inside the residual bodies of the rat blastocyst, the presence of an exogenous marker (horseradish peroxidase) has been observed (Schlafke and Enders, 1973; Dvořák and

Trávník, 1975). Schlafke and Enders (1967) state that in the formation of these "degradation bodies" both phagocyted material and products of autolysis can participate. Acid phosphatase activity (Šťastná 1977a) has been found in almost all residual bodies.

Smooth Endoplasmic Reticulum

A structural form of smooth endoplasmic reticulum such as that described for the earlier stages has not been observed at the blastocyst stage. Small vesicles account for 0.5% of the cytoplasmic volume and are distributed individually in cytoplasmic zones containing cell organelles whose character is difficult to define and all of which belong to smooth endoplasmic reticulum (Dvořák et al., 1977). Smooth vesicles in the vicinity of the nucleus mostly contained organophosphate-sensitive esterase activity and the vesicles near the cell membrane organophosphate-resistant esterase activity (Trávník, 1977b, 1978).

Granular Endoplasmic Reticulum

Granular endoplasmic reticulum is located in zones containing cytoplasmic organelles (Figs. 45 and 46). It has the form of single, sometimes branching, unevenly arranged cisternae, provided with ribosomes at regular distances. Some of the cisternae are slightly dilated and contain medium electron dense amorphous material, and others are attached close to mitochondria (see above). Direct connection of some cisternae with the nuclear envelope was observed by Enders and Schlafke (1967). Schlafke and Enders (1967) state that granular endoplasmic reticulum at this stage is present in greater quantities than the preceding stages. This agrees with morphometric findings (Dvořák et al., 1977) where the volume of granular endoplasmic reticulum is 0.25% of the volume of the ovum cytoplasm. In granular endoplasmic reticulum, acid phosphatase (Šťastná, 1977a) and organophosphate-sensitive esterase activity (Trávník, 1977b, 1978) is present.

Ribosomes

In addition to ribosomes, which are part of granular endoplasmic reticulum or attached to the outer membrane of the nuclear envelope, there are many free ribosomes in the cytoplasm of embryoblast cells (Figs. 45 and 46). Ribosomes arranged in polysomes occur mostly in cytoplasmic zones containing cell organelles. The regions of ground cytoplasm filled with ribosomes are larger than those of the blastomeres of the earlier stages. Schlafke and Enders (1967) state that a great number of polysomes are present.

Golgi Complex

The Golgi complex (Fig. 45) accounts for a negligible part of the cytoplasm in embryoblast cells. It is located, often in the form of several separated dictyosomes, in the perinuclear region. Individual dictyosomes consist of four to eight cisternae running parallely with the nuclear envelope.Small smooth vesicles located at the sides of parallely ordered cisternae and on both surfaces of dictyosomes, but in the greatest amount on the surface usually turned away from the nucleus, are a regular part ot the Golgi complex. The vesicles have a variable appearance; some of them contain amorphous material of different electron density. Both cisternae and vesicles of the Golgi complex show acid phosphatase activity (Šťastná, 1977a, 1978a).

Centrioles

Centrioles have not been observed in our material.

Lamellar Structures

Lamellar structures (Fig. 44) fill a considerable part of the cytoplasm of embryoblast cells, taking up 23% of the cytoplasmic volume (Dvořák et al., 1977). When compared to the earlier stages, the lamellar structures are, as a rule, shorter in the ultrathin section, and areas with parallely arranged lamellae are of smaller volume. In many regions, degradation of lamellar structures is distinctly evident (Fig. 46). Organophosphate-sensitive esterase activity (Trávník, 1977b, 1978) has also been found at this stage on lamellar structures.

Glycogen

The stage of the early blastocyst, by nature of its distribution character and the morphology of glycogen particles, greatly resembles the relationships in the eight-cell ova (Fig. 47). The frequency of occurrence of glycogen particles in the cytoplasm of the cells of the early blastocyst is approximately the same. More conspicuous differences between the cells of the embryoblast and the trophoblast have not been observed.

Glycogen particles show considerable variability in size and are usually not distributed in the cytoplasm very evenly. The presence of smooth vesicles with glycogen is no longer to be found. Autophagic vacuoles, unlike those of the eight-cell ovum, were more or less filled with glycogen over the whole area. In some autophagic vacuoles, glycogen still retained its granular appearance, while in others, glycogen particles coalesced into large compact clusters. Individual particles and/or their aggregations were regularly contained in residual bodies. Extracellular location of glycogen is comparatively frequent. Aggregations of glycogen particles occur in intercellular spaces, mainly near the free surface of embryoblast cells, partly limiting the developing blastocyst cavity.

Lipid Droplets

Lipid droplets were observed only occasionally, mostly in the cortical cytoplasm of embryoblast cells. Sometimes, they were even located inside autophagic vacuoles; they comprise 0.7% of the cytoplasmic volume (Dvořák et al., 1977).

Cell Membrane

Embryoblast cells are mostly attached to each other; sometimes, they are separated by a narrow intercellular space into which the neighboring cells project by short irregular microvilli. Tonofilaments have not been observed in the cytoplasm of the embryoblast or trophoblast cells. The surface of embryoblast cells turned into the lumen of the blastocyst is more or less smooth or provided with short irregular microvilli. The contact of the cells of the embryoblast shows no differentiated junctional structures; a structure resembling a close junction is only found in places. In some blastocysts, alkaline phosphatase activity (Šťastná, 1979) has been found on the membranes of embryoblast cells.

2.6.2. Trophoblast Cells

Trophoblast cells are located in one layer below the zone pellucida and consist of a nucleus and cytoplasm. The size of trophoblast cells is rather variable due to their irregular shape.

2.6.2.1. Nucleus

The nuclei (Fig. 43) of trophoblast cells are, depending on the shape of the cells, oval to considerably flattened. The average size of the nuclei is 5 μm. Their surface is sometimes smooth, but often minor irregularities to deep invaginations can be observed. This characteristic distinguishes them from the nuclei of embryoblast cells.

Nuclear Envelope

The regular width of the nuclear envelope consisting of two membranes is 15 - 20 nm. It contains numerous pores usually filled with electron dense filamentous material. In isolated cases, intranuclear annulate lamellae were observed in the nucleus. Ribosomes join the outer membrane at irregular distances. In the nuclear envelope of trophoblast cells, acid phosphatase activity (Šťastná, 1977a) has been found regularly, and in some cells, organophosphate-sensitive esterase activity (Trávník, 1977b, 1978) has also been observed.

Chromatin
The distribution of chromatin in the nucleus is regular and in places slightly condensed.

Nucleolus

The nuclei of trophoblast cells contain one or two nucleoli consisting of the voluminous pars granulosa of reticular arrangement, mostly a small central region of homogenous appearance, i.e., the pars fibrosa.

2.6.2.2. Cytoplasm

As is the case embryoblast cells, organelles of trophoblast cells are segregated in the perinuclear and cortical zones of cytoplasm and connecting columns (Fig. 48). The remaining parts of ground cytoplasm are filled with lamellar structures. Although lamellar structures still fill large areas of cytoplasm, the areas are not as distinctly delineated as in previous stages.

Mitochondria

Mitochondria (Fig. 48) amount to 4.0% of the cytoplasmic volume (Dvořák et al., 1977). They are located in cytoplasm together with ribosomes and other cell organelles. By nature of their submicroscopic structure and close relation to granular endoplasmic reticulum, they are quite similar to the mitochondria of embryoblast cells. In some mitochondria, succinate dehydrogenase activity (Šťastná, 1977b) has been proved; occasionally organophosphate-resistant esterase activity (Trávník, 1977b, 1978) has also been found on mitochondrial cristae.

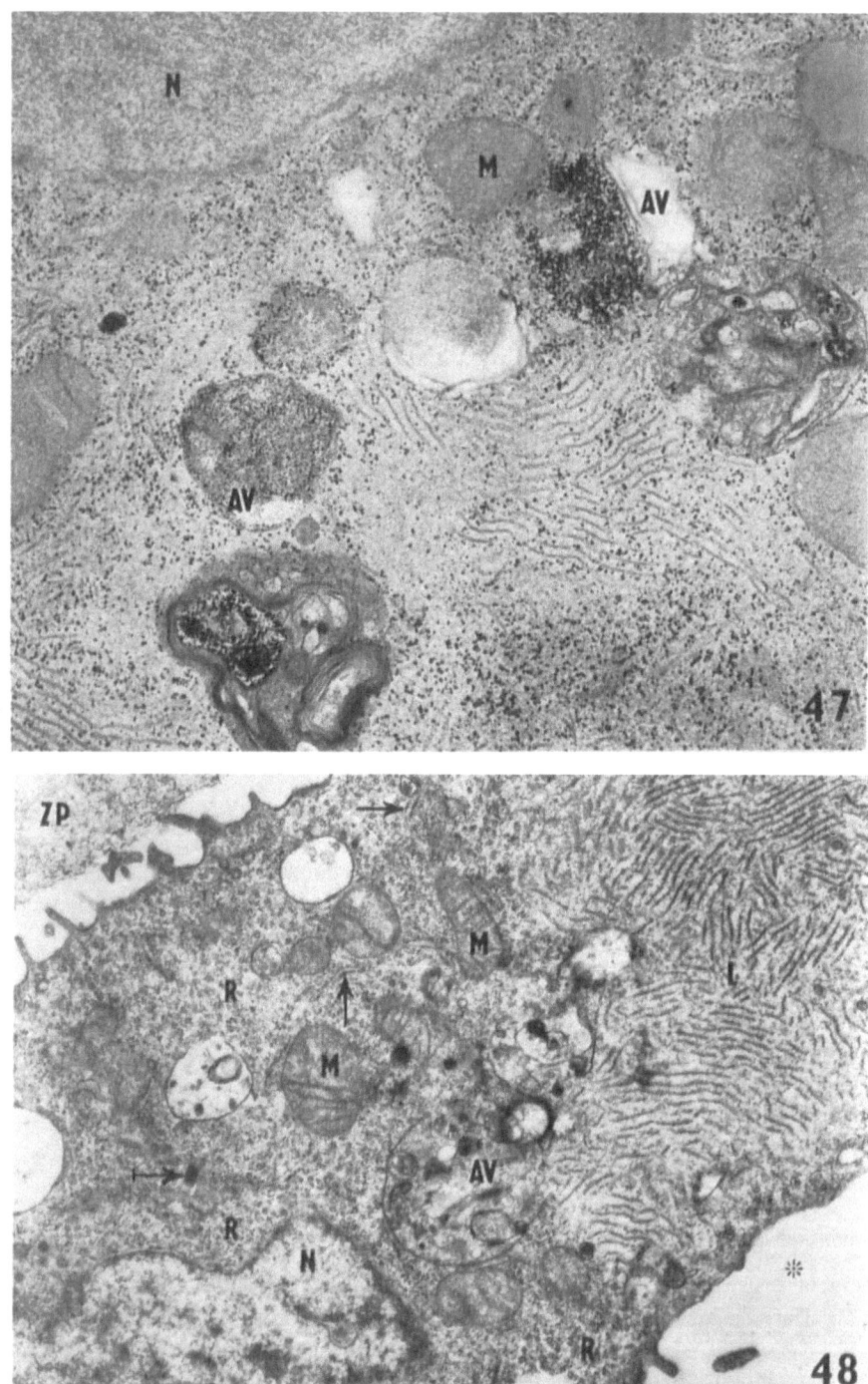

Figs. 47 and 48

Fig. 47. Early rat blastocyst. Demonstration of glycogen particles. Nucleus (*N*); autophagic vacuoles containing more or less of glycogen (*AV*); mitochondria (*M*). Fixation glutaraldehyde and OsO$_4$; embedding medium Durcupan ACM; nonstained ultrathin section; magnification X 26,000

Autophagic Vacuoles

Autophagic vacuoles (Figs. 48 and 49) are located in zones containing organelles in trophoblast cells. Their quantitative occurrence is variable. Due to their morphologic properties and the presence of acid phosphatase (Šťastná 1977a) and organophosphate-resistant esterase activity (Trávník, 1977b, 1978), they are quite similar to autophagic vacuoles in the cells of the embryoblast.

Residual Bodies

Residual bodies (Fig. 49) in trophoblast cells are represented at approximately the same rate as in embryoblast cells. Together with autophagic vacuoles, they account for about 3.0% of the cytoplasmic volume (Dvořák et al., 1977). Residual bodies are often aggregated in high numbers in trophoblast cells; together with lipid droplets, they occupy large regions of the cytoplasm. Acid phosphatase activity has been found in residual bodies (Šťastná, 1977a).

Smooth Endoplasmic Reticulum

Smooth endoplasmic reticulum analogous to earlier stages of development has not been observed. Parts of it may be smooth vesicles distributed in ground cytoplasm, particularly in regions filled with ribosomes and other cell organelles. They become visible after marking pinocytotic vesicles with exogenous proteins, using horseradish peroxidase (Fig. 50). Quantitative measurements determined its amount to be 0.2% of the cytoplasmic volume (Dvořák et al., 1977). In smooth vesicles in the vicinity of the nucleus, organophosphate-sensitive esterase activity was observed and in vesicles located near the cell membrane, organophosphate-resistant esterase activity (Trávník, 1977b, 1978) was found.

Granular Endoplasmic Reticulum

As can be observed in embryoblast cells, granular endoplasmic reticulum has the form of individually distributed cisternae located in the regions with cell organelles (Fig. 48). Some of its cisternae are closely related to mitochondria. Its volume is 0.4% of the cytoplasmic volume (Dvořák et al., 1977). In part of the cisternae of granular endoplasmic reticulum, organophosphate-sensitive esterase activity (Trávník, 1977b, 1978) and acid phosphatase activity (Šťastná, 1977a) has been proved.

Ribosomes

Ribosomes, in the form of polysomes, fill ground cytoplasm in the zones with organelles (Fig. 48). Part of them are bound to granular endoplasmic reticulum and the outer membrane of the nuclear envelope.

Fig. 48. Early rat blastocyst. Part of a trophoblast cell with nucleus (*N*), mitochondria (*M*), autophagic vacuole (*A V*), ribosomes (*R*), granular endoplasmic reticulum (→), and centriole (→); lamellar structures (*L*); zona pellucida (*ZP*); blastocyst cavity (∗). Fixation glutaraldehyde and OsO_4; embedding medium Durcupan ACM; magnification X 17,000

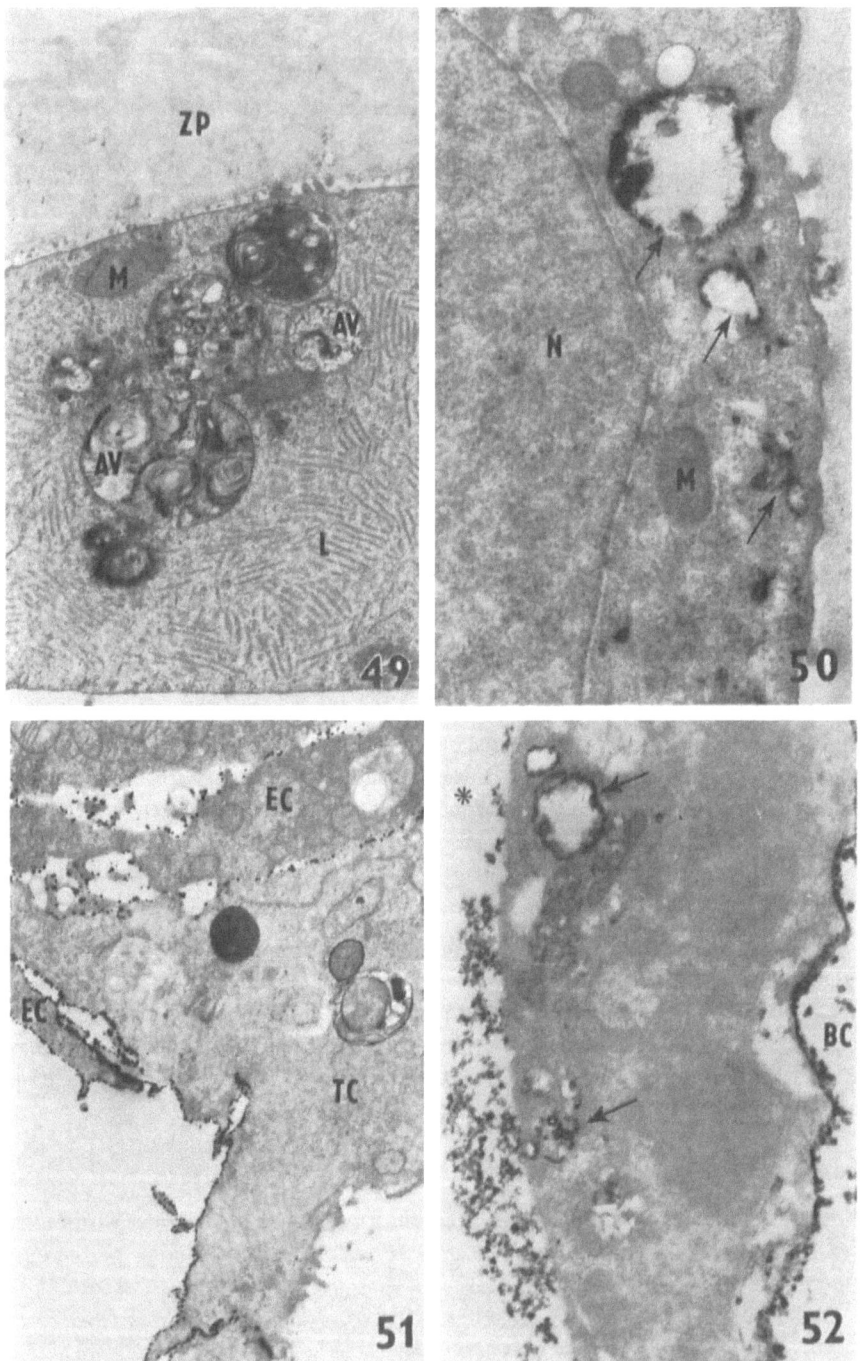

Figs. 49–52

Fig. 49. Early rat blastocyst. Part of a trophoblast cell cytoplasm with a group of autophagic vacuoles (AV); mitochondrion (M); lamellar structures (L); zona pellucida (ZP). Fixation glutaraldehyde and OsO₄; embedding medium Durcupan ACM; magnification X 14,000

Golgi Complex

The Golgi complex (Fig. 48), in the form of several independent dictyosomes, occurs in the perinuclear region of trophoblast cells. Their submicroscopic structure is very similar to the structure in the dictyosomes of embryoblast cells. The Golgi complex accounts for only a negligible part of cytoplasmic volume. An associated granule is only occasionally a compartment of the Golgi complex (Schlafke and Enders, 1967). Acid phosphatase activity (Šťastná, 1977a, 1978a) was found both in cisternae and vacuoles of the Golgi complex.

Centrioles

Centrioles have not been observed in trophoblast cells.

Lamellar Structures

Lamellar structures (Figs. 48 and 49) fill 14% of the cytoplasmic volume (Dvořák et al., 1977). Their morphologic features do not differ substantially from those described for embryoblast cells. Small islets of material showing organophosphate-resistant esterase activity have been found in close contact with lamellar structures (Trávník, 1977b, 1978).

Glycogen

The frequency of occurrence of glycogen particles (β-type) in trophoblast cells cannot be distinguished from the frequency in the cells of the embryoblast by a simple estimate. Glycogen particles are of smaller size, and in the cytoplasm of trophoblast cells, they are mostly regularly distributed. As is the case in the embryoblast, they are always numerous inside autophagic vacuoles, individually and/or as small clusters also in residual bodies. Glycogen particles have also been demonstrated at the trophoblast cells in the blastocyst cavity.

Lipid Droplets

Lipid droplets are located in trophoblast cells in a greater amount than in the cells of the embryoblast, either individually or often in aggregations. They are chiefly found in trophoblast cells situated at the abembryonal pole. Lipid droplets have often been observed inside residual bodies.

Fig. 50. Early rat blastocyst. Reaction for exogenous microperoxidase. A part of trophoblast cell with nucleus (*N*), mitochondria (*M*); reaction product (→) is present in numerous smooth vesicles of different size. Fixation glutaraldehyde and OsO_4; embedding medium Durcupan ACM; nonstained ultrathin section; magnification X 20,000. From Dvořák, M., Trávník, P., Histochemistry 47, 257–262 (1976)

Fig. 51. Early rat blastocyst. Demonstration of alkaline phosphatase in the trophoblast cell (*TC*) and embryoblast cell (*EC*). Fixation glutaraldehyde and OsO_4; embedding medium Durcupan ACM; nonstained ultrathin section; magnification X 11,500

Fig. 52. Early rat blastocyst. Reaction for exogenous horseradish peroxidase. A part of a trophoblast cell with numerous smooth vesicles of different size with reaction product (→), which is also present at the cell surfaces of perivitelline space (∗) and blastocyst cavity (*BC*). Fixation glutaraldehyde and OsO_4; embedding medium Durcupan ACM; nonstained ultrathin section; magnification X 16,500

Cell Membrane

Membranes of the free surface of trophoblast cells facing the zona pellucida are greatly indented. They project into numerous long irregular microvilli located at irregular distances (Fig. 48). Schlafke and Enders (1967) state that microvilli are generally longer and have a more regular shape than in previous stages. In some membranes, signs of pinocytotic activity are clearly visible. The surfaces of the neighboring trophoblast cells are also greatly indented, with numerous microvilli projecting into the intercellular space. Sometimes microvilli of the neighboring cells are in close contact. In the outer part of the trophoblast cell contact at the perivitelline space, zonulae occludentes are present in the early blastocyst. For the remaining part of the contact interdigitations of cytoplasmic processes of the adjacent cells are characteristic. A structure corresponding to the zonula adherens is situated below it, but not always. In a single case out of several hundreds of structures observed, it was possible to evaluate the junctional structures as a desmosome.

There is a narrow intercellular space between the cells of the embryoblast and trophoblast; microvilli occur more frequently in the early blastocyst than in the late blastocyst. Some small junctions between the cells correspond to the close junction type and in isolated cases also to the zonula adherens. The glycocalyx is rendered visible on the free surface of trophoblast cells. In some cases, material on the surfaces between the cells of the trophoblast, between the cells of the trophoblast and the embryoblast, and between the cells of the embryoblast could be seen after staining with ruthenium red. Alkaline phosphatase activity (Fig. 51) has been found (Šťastná, 1979) at the membranes of the trophoblast cells. Some sections of the cell membrane showed organophosphate-sensitive esterase activity (Trávník, 1977b, 1978). Exposure of the blastocyst to the influence of horseradish peroxidase causes the penetration of this protein into the cytoplasm and from there into the blastocyst cavity (Fig. 52).

2.6.3. Zona Pellucida

The zona pellucida is present only in the early 5-day-old blastocyst. It is stained to a smaller extent with ruthenium red than the eight-cell ovum. The zona pellucida began to lyse at 2:00 p.m.; at 4:00 p.m., 46% was lysed and at 8:00 p.m., all ova were naked (Surani, 1975).

2.7. Late Blastocyst

Late blastocysts (Fig. 53) were obtained at 6:00 p.m. - 8:00 p.m. on the 5th day or at 6:00 a.m. on the 6th day of pregnancy. The longitudinal axis passes through the embryoblast and the abembryonal pole. All blastocysts lacked the zona pellucida, and such ova possessed a characteristic shape. After leaving the zona pellucida, the blastocyst enlarges slightly and becomes more elongated (Schlafke and Enders, 1967). At the same time, the blastocyst cavity takes on a round shape. The average size of the late blastocyst is 60 - 85 μm.

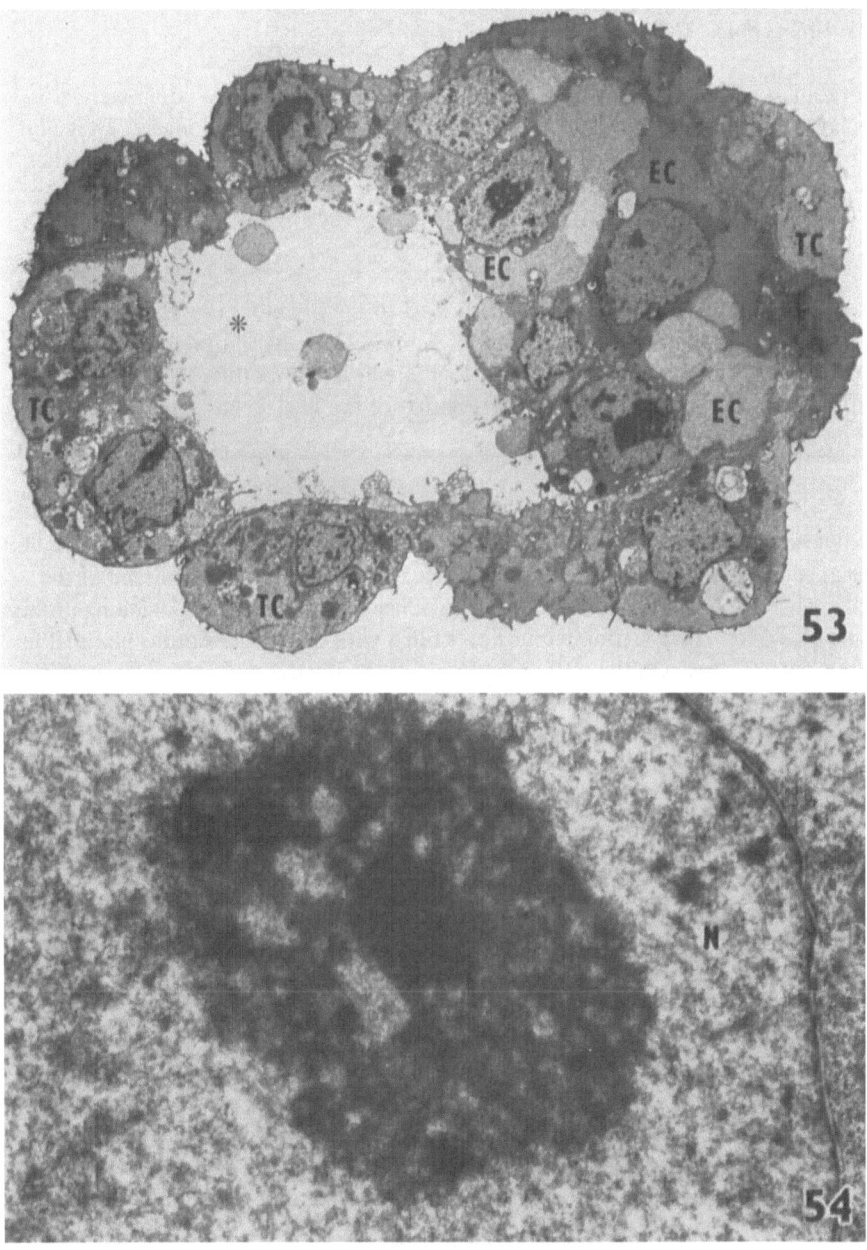

Fig. 53. Late rat blastocyst. Trophoblast cells (*TC*); embryoblast cells (*EC*); blastocyst cavity (∗).
Fixation glutaraldehyde and OsO_4; embedding medium Durcupan ACM; magnification X 2,700

Fig. 54. Late rat blastocyst. Nucleus (*N*) of an embryoblast cell, with a distinct reticular nucleolus.
Fixation glutaraldehyde and OsO_4; embedding medium Durcupan ACM; magnification X 20,000

2.7.1. Embryoblast Cells

Embryoblast cells are of polygonal shape and 11.5 μm in size. They are mostly closely attached to each other, and large intercellular spaces can be seen in places. They consist of the nucleus and the cytoplasm in which signs of segregation of cell organelles are visible even at this stage.

2.7.1.1. Nucleus

The nuclei of embryoblast cells are mostly oval to slightly elongated (Fig. 53). They are located approximately in the center of embryoblast cells and their diameter is about 5.5 μm. The nucleus surface is either more or less smooth or exhibits small, irregularly distributed folds. The nucleus consists of the nuclear envelope, chromatin, and nucleoli.

Nuclear Envelope

The nuclear envelope consists of two membranes running parallel which form the perinuclear space with a width of 15 - 20 nm (Fig. 54). On the outer membrane of the nuclear envelope of some embryoblast cells, conspicuous signs of the blebbing process can be observed. There are numerous pores filled with dense filamentous material in the nuclear envelope. On the outer membrane of the nuclear envelope ribosomes are attached at irregular distances. Intranuclear annulate lamellae were no longer observed in the nuclei of embryoblast cells at this stage. Acid phosphatase (Šťastná, 1977a) was proved on the nuclear envelope.

Chromatin

Nuclear chromatin in the nucleus of embryoblast cells is, in the interphase, either regularly scattered or forms typical small karyosomes, especially near the nuclear envelope or less frequently in other places of the nucleus.

Nucleolus

In the nucleus one to two nucleoli of reticular structure are located in which the pars granulosa and pars amorpha prevail over the pars fibrosa (Fig. 54). The size of the nucleoli varies from 2 - 3 μm. Dvořák (1974a) states that most of the nucleoli (85.7%) at this stage have no compact centers. Tiny bodies of electron dense material similar in appearance to the substance of the nucleolus can often be seen between the nucleolus and the nuclear envelope. The location of the nucleoli is variable. Most often, they are located centrally in the nucleus; sometimes, however, they are found near the nuclear envelope. At this stage, well-developed perinucleolar chromatin was observed for the first time in the course of cleavage around the nucleolus.

2.7.1.2. Cytoplasm

In the cytoplasm of embryoblast cells, the segregation of cell organelles into the perinuclear and the peripheral zones and columns connecting the two zones is still clearly expressed (Fig. 55). The zones containing cell organelles, unlike in the earlier stages, prevail over regions of ground cytoplasm filled with lamellar structures. In this stage, the lamellar structures are in an advanced state of degradation, so that one can often observe large regions filled with fine amorphous or cloddy, medium electron dense material in which remnants of lamellar structures are located.

Mitochondria

Mitochondria form 4.7% of the cytoplasmic volume (Dvořák et al., 1977). Unlike in the earlier stage, mitochondria of the embryoblast cells are of regular rod-like to filament-like elongated shape (Fig. 56); their length varies most frequently between 2 and 4 μm (but they can reach lengths over 5 μm), and their width is about 0.2 μm. Very often, constricted or branched mitochondria can be found. Mitochondrial cristae are more frequent, oriented mostly perpendicular to the longitudinal axis of the mitochondrion, often partitioning the mitochondrion. In isolated cases, arch-like cristae and/or cristae composed of three membranes, as were described in the earlier stages, still occur. The density of the mitochondrial matrix is greatly variable. Cisternae of granular endoplasmic reticulum, which lack ribosomes on the side turned toward the mitochondria are regularly attached to mitochondria. Several mitochondria are often attached to one cisterna. In the mitochondria of embryoblast cells of some blastocysts, a weakly positive succinate dehydrogenase activity (Šťastná, 1977b) has been found.

Autophagic Vacuoles

Autophagic vacuoles (Fig. 56) are found in the embryoblast cells in a relatively low number. They reach a size of 1 - 2 μm and are mostly limited by one membrane; autophagic vacuoles with two membranes were only observed in isolated cases. These, probably initial, forms of autophagic vacuoles frequently contained ground cytoplasm with ribosomes or lamellar structures, sometimes even with small vesicles, but no other cell organelles were observed. The appearance of autophagic vacuoles with morphologically apparent signs of digestion is greatly variable. The autophagic vacuoles often appear to be light and contain only a small amount of fine flake-like material, tiny vesicles, and particles of electron denser amorphous material. In most of the autophagic vacuoles, acid phosphatase activity has been proved (Šťastná, 1977a) as well as that of nonspecific esterase (Trávník, 1977b, 1978).

Residual Bodies

Residual bodies are located in zones containing organelles. Together with autophagic vacuoles, they take up 2.8% of the cytoplasmic volume (Dvořák et al., 1977), their representation being approximately the same as in the earlier stage. They are limited by a single membrane and characterized by a high content of membrane fragments and myeline figures and different amounts of amorphous dense material. They show positive reaction to acid phosphatase (Šťastná, 1977a) and to nonspecific esterase (Trávník, 1977b, 1978).

Smooth Endoplasmic Reticulum

Smooth endoplasmic reticulum consists in the cytoplasm of embryoblast cells of a low number of individually scattered smooth vesicles, their total volume corresponding, on the average, to 0.45% of the cytoplasmic volume (Dvořák et al., 1977). In smooth vesicles located near the nucleus, organophosphate-sensitive esterase activity (Trávník, 1977b, 1978) was found.

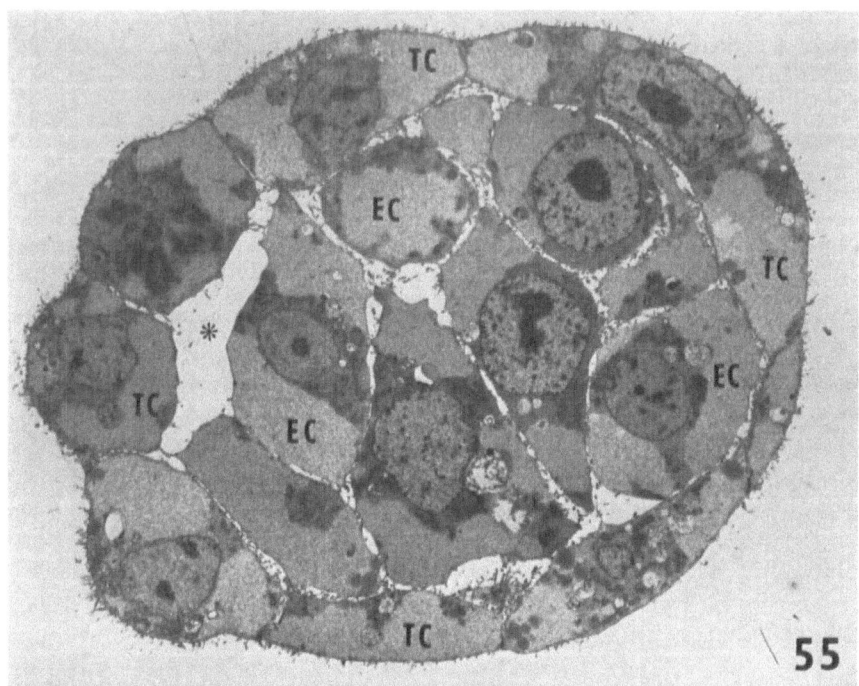

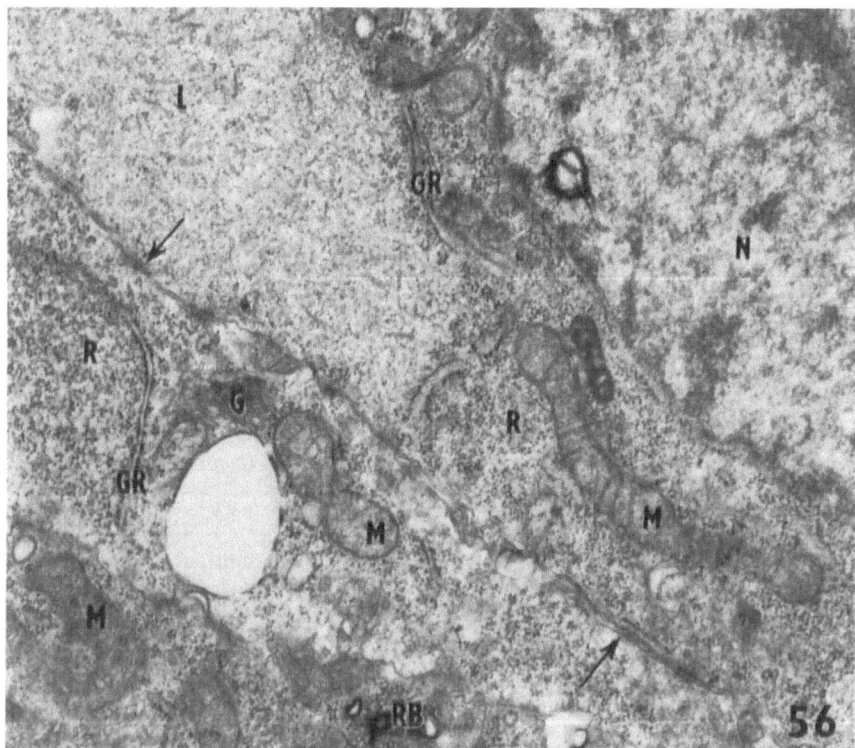

Figs. 55 and 56

Granular Endoplasmic Reticulum

Granular endoplasmic reticulum is located in the cytoplasm in the form of single long and often branched cisternae (Fig. 56), a part of which are characterized by their close relation to mitochondria. Its total amount has risen in comparison with the early blastocyst, forming 0.97% of the cytoplasmic volume (Dvořák et al., 1977). Cisternae of granular endoplasmic reticulum are often dilated and filled with medium electron dense flake-like material. Acid phosphatase activity (Šťastná, 1977a) was found in cisternae of granular endoplasmic reticulum.

Ribosomes

The amount of free ribosomes (Fig. 56) has increased in comparison to the preceding stage. They fill ground cytoplasm of the regions containing organelles and have the form of polysomes. Bound ribosomes are connected to the structures of granular endoplasmic reticulum or attached to the outer membrane of the nuclear envelope.

Golgi Complex

The total volume of the Golgi complex in the embryoblast cells of late blastocyst has substantially increased in comparison to the preceding stage. It comprises 0.24% of the cytoplasmic volume (Dvořák et al., 1977). It is divided into several separated dictyosomes located in the perinuclear zone of the cytoplasm. Compared with the earlier stage, within the individual fields of the Golgi complex, the number of cisternae and vesicles increases. Acid phosphatase activity (Šťastná, 1977a, 1978a) was found in both.

Centrioles

Centrioles have not been observed in embryoblast cells.

Lamellar Structures

The quantity of lamellar structures has decreased in embryoblast cells at this stage and account for 17% of the cytoplasmic volume (Dvořák et al., 1977). They are very short and irregularly arranged. In many areas, they are evidently degraded; those zones are then filled with reticular irregularly arranged material in which fragments of lamellar structures are located (Fig. 56). Traces of organophosphate-resistant esterase activity (Trávník, 1977b, 1978) were found on lamellar structures.

Glycogen

In late blastocysts, glycogen is present in a minimum quantity. Some cells still contain a small amount, while in others it is no longer present. Glycogen occurs in ground cy-

---

Fig. 55. Late rat blastocyst. Ultrathin section cut in a plane nearly out of blastocyst cavity (∗); several embryoblast cells (*EC*) and trophoblast cells (*TC*). Fixation glutaraldehyde and OsO₄; embedding medium Durcupan ACM; magnification X 2,000

Fig. 56. Late rat blastocyst. Contact of two embryoblast cells (→); nucleus (*N*); mitochondria (*M*); granular endoplasmic reticulum (*GR*); ribosomes (*R*); Golgi complex (*G*); residual body (*RB*), remnants of lamellar structures (*L*). Fixation glutaraldehyde and OsO₄; embedding medium Durcupan ACM; magnification X 30,000

toplasm, in which it is scattered into infrequent very fine particles, smaller in sizes than the typical β-particles. Rarely, traces of glycogen can also be found in some autophagic vacuoles and residual bodies.

Lipid Droplets

In embryoblast cells, lipid droplets occur individually, their total volume being 1.8% of the cytoplasmic volume (Dvořák et al., 1977).

Cell Membrane

Embryoblast cells are mostly closely attached to each another, and intercellular spaces are quite narrow with a regular width. In some places, the surface of embryoblast cells is uneven, and the width of the intercellular space is then irregular; the cells project into it by short irregular microvilli. The connection of cells is strengthened either by primitive desmosomes of the same structure and arrangement as in the preceding stage or by means of the zonulae occludentes. The connection between embryoblast and trophoblast cells is arranged in the same way.

2.7.2. Trophoblast Cells

Trophoblast cells are flat to cuboidal in shape (Figs. 53 and 55) and consist of a nucleus and cytoplasm.

2.7.2.1. Nucleus

The shape of the nucleus is mostly oval to flattened depending on the shape of the trophoblast cell. More often than in the nuclei of embryoblast cells, irregular, frequently even rather deep invaginations on the nuclear surface are found (Fig. 57). The average size of trophoblast cell nuclei is 8 μm in length and 5 μm in width.

Nuclear Envelope

The nuclear envelope has the usual structure (Fig. 58). It is formed by two membranes which form the perinuclear space with a regular width of 15 - 20 nm. It contains numerous nuclear pores filled with amorphous electron dense material. Ribosomes are attached to the outer membrane of the nuclear envelope at irregular distances. Intranuclear annulate lamellae have not been observed. Acid phosphatase activity (Šťastná, 1977a) has been proved in the nuclear envelope (Fig. 59).

Chromatin

Nuclear chromatin is distributed regularly in the nucleus, rendering the nuclei light, or it forms small karyosomes, particularly at the nuclear envelope.

Fig. 57. Late rat blastocyst. Trophoblast cell with nucleus (N), mitochondria (M), Golgi complex (G), and remnants of lamellar structures (L). Fixation glutaraldehyde and OsO_4; embedding medium Durcupan ACM; magnification X 20,000

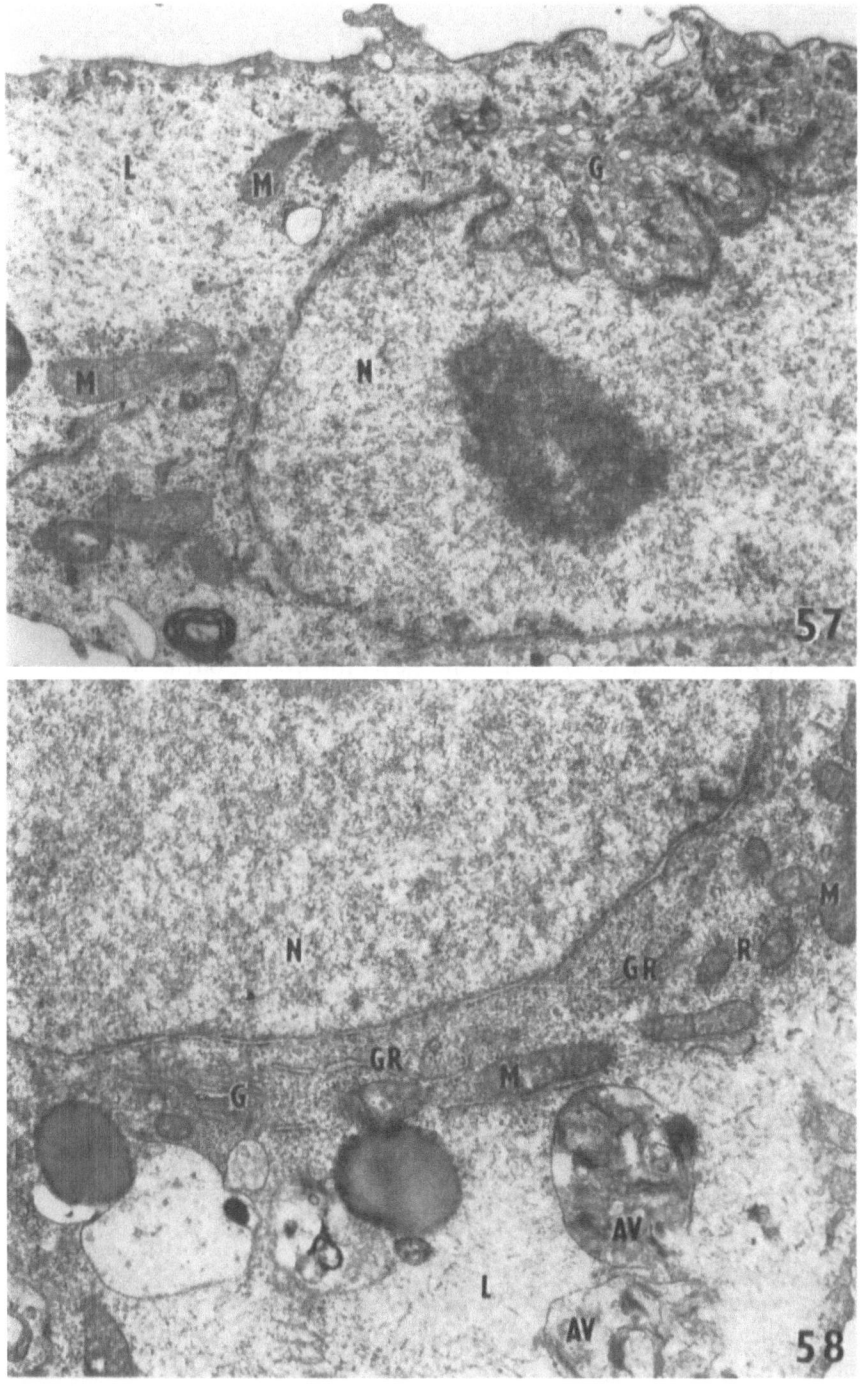

Figs. 57 and 58

Fig. 58. Late rat blastocyst. Trophoblast cell with nucleus (N), Golgi complex (G), mitochondria (M), autophagic vacuoles (AV), granular endoplasmic reticulum (GR), ribosomes (R), and remants of lamellar structures (L). Fixation glutaraldehyde and OsO$_4$; embedding medium Durcupan ACM; magnification X 20,000

Nucleolus

There are one or two nucleoli of already typical reticular structure in the trophoblast cell nucleus (Figs. 57 and 59). Dvořák (1974a) states that most of the nucleoli (85.7%) at this stage no longer have compact centers. In some blastocysts, the nucleoli of trophoblast cells are surrounded by conspicuous, morphologically fully developed perinucleolar chromatin.

2.7.2.2. Cytoplasm

The segregation of cell organelles is still visible in cytoplasm. Zones containing lamellar structures are, however, mostly filled with only fine amorphous material of medium electron density (Fig. 58), in which fragments of lamellar structures are scattered. In individual samples of this stage, the degradation degree of lamellar structures is fairly variable. Enders and Schlafke (1967) state that areas which are largely filled by the fibrous plaques can still be seen, although the diameter of these plaques appears somewhat reduced and the margins of the regions are not always as distinct as in earlier stages.

Mitochondria

Mitochondria account for 4.3% of the cytoplasmic volume (Dvořák et al., 1977). They are located in zones containing organelles and are mostly oval to rod-like in shape; filamentous mitochondria are only rarely present. Their length in section as a rule does not exceed 2 μm, their width being about 0.2 - 0.4 μm. As is the case in embryoblast cells, a close spatial relationship of mitochondria to cisternae of granular endoplasmic reticulum is characteristic of trophoblast cells. In the mitochondria of trophoblast cells of some blastocysts, a weak succinate dehydrogenase activity (Šťastná, 1977b) has been proved.

Autophagic Vacuoles

Newly developed autophagic vacuoles containing identifiable organelles only occur in isolated cases (Fig. 58). Autophagic vacuoles with morphologically apparent signs of digestion of the content have structures similar to those described in earlier stages. Often, autophagic vacuoles have been observed in which lamellar structures were separated. Most autophagic vacuoles have shown positive reactions to acid phosphatase activity (Šťastná, 1977a) and organophosphate-resistant esterase activity (Trávník, 1977b, 1978).

Residual Bodies

Residual bodies located in zones containing cell organelles comprise, together with autophagic vacuoles, 4.7% of the cytoplasmic volume (Dvořák et al., 1977), which means that these structures occur two times more frequently in trophoblast cells than in embryoblast cells. The appearance of residual bodies is variable; they often resemble dense bodies described in various kinds of somatic cells. They mostly show acid phosphatase activity (Šťastná, 1977a) so that they can be ranked among the secondary lysosomes.

Smooth Endoplasmic Reticulum

Smooth endoplasmic reticulum is in no case a prominent cytoplasmic structure; it amounts to only 0.65% of the cytoplasmic volume (Dvořák et al., 1977). It has the form of individually scattered smooth vesicles. In the smooth vesicles, particularly in the vicinity of the cell membrane, organophosphate-resistant esterase activity was present (Trávník, 1977b, 1978).

Granular Endoplasmic Reticulum

Granular endoplasmic reticulum (Fig. 58) is located in zones of cytoplasm containing cell organelles. Compared to embryoblast cells of the same stage, it is present in a higher amount (1.5% of the cytoplasmic volume — Dvořák et al., 1977). It has the shape of mostly only short, isolated cisternae. In the cisternae of granular endoplasmic reticulum, acid phosphatase activity (Šťastná, 1977a) has been proved.

Ribosomes

Ribosomes are located in zones containing cell organelles and have the form of polysomes (Fig. 60). The areas of ground cytoplasm filled with polysomes are smaller than those in embryoblast cells. Part of the total number of ribosomes are bound to membranes of granular endoplasmic reticulum and the outer membrane of the nuclear envelope.

Golgi Complex

As is the case in embryoblast cells, the Golgi complex (Fig. 60) is located in the perinuclear region in the form of several Golgi fields, consisting of six to eight cisternae oriented parallel to the nuclear envelope and of a variable amount of vesicular structures. The total volume of the structures of the Golgi complex in trophoblast cells in comparison to the early blastocyst is markedly increased, forming 0.47% of the cytoplasmic volume (Dvořák et al., 1977). In the structures of the Golgi complex of trophoblast cells, acid phosphatase activity (Šťastná, 1977a, 1978a) was present.

Centrioles

Centrioles were found in trophoblast cells in our material in only three cases. They occurred — two in one cell — near the nuclear envelope, quite distant from each other. They are oriented approximately at right angles.

Lamellar Structures

The total volume of lamellar structures is smaller than in the early blastocyst, taking up only 11% of the cytoplasmic volume (Dvořák et al., 1977). As stated above, lamellar structures are fragmented in many cases, and often only disarranged remnants remain (Figs. 58 and 60). Traces of organophosphate-resistant esterase activity (Trávník, 1977b) were found on lamellar structures.

Glycogen

The structure and distribution of glycogen in trophoblast cells are the same as in embryoblast cells.

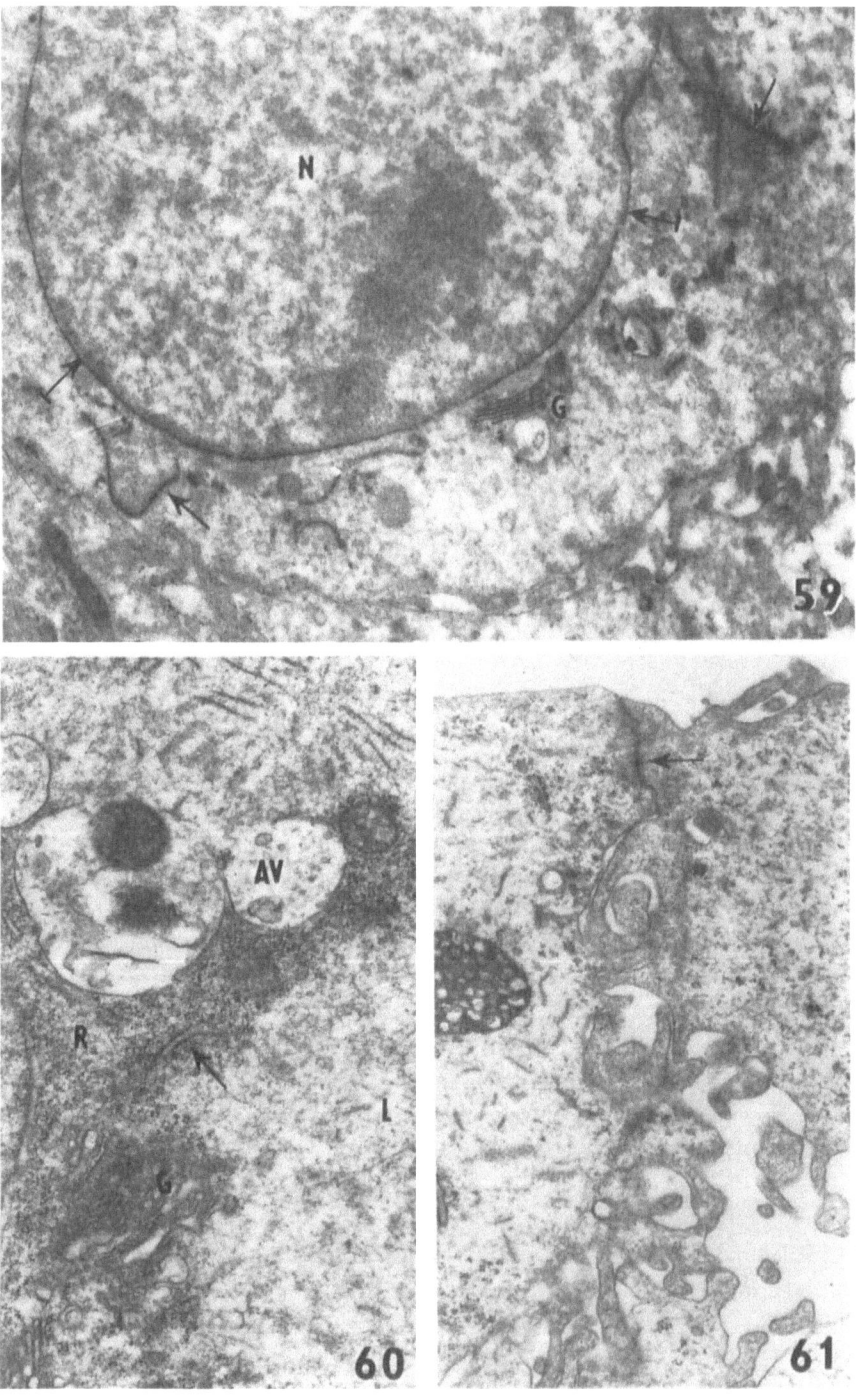

Figs. 59—61

Fig. 59. Late rat blastocyst. Demonstration of acid phosphatase activity in trophoblast cell in Golgi complex (G), endoplasmic reticulum (→) and nuclear envelope (→); nucleus (N). Fixation glutaraldehyde and OsO₄; embedding medium Durcupan ACM; nonstained ultrathin section; magnification X 20,000. From Šťastná, J., Scripta med. 50, 21—34 (1977)

Lipid Droplets

The total amount of lipid droplets (Fig. 58) in comparison to that in the early blastocyst has increased to 5.0% of the cytoplasmic volume (Dvořák et al., 1977). They are located in cytoplasm individually or in large groups. Lipid droplets are chiefly found in zones of cytoplasm containing organelles.

Cell Membrane

The external free surface of trophoblast cells is much indented, projecting in long irregular microvilli which are mostly simple (Fig. 61) but sometimes branched. The cell membranes of the adjacent trophoblast cells are also much indented. At their outer surfaces, trophoblast cells are firmly connected by means of junctional complexes (Fig. 61). On some projections of adjacent cells touching each other, small desmosomes of the primitive type and of the same structure as in the earlier stages are sometimes formed. At this stage, true desmosomes with developed tonofilaments were observed for the first time between trophoblast cells. Fine bundles of tonofilaments were also observed in isolated cases elsewhere in the cytoplasm of trophoblast cells. The glycocalys is stained exclusively on the free surface of the ovum, caused by the fact that ruthenium red does not affect intercellular spaces. Microvilli occur on the outer surface of the cells of the trophoblast less frequently than in the early blastocyst. On the outer side of the cell membrane, organophosphate-sensitive esterase activity (Trávník, 1977b, 1978) was proved at this stage.

2.7.3. Zona Pellucida

The zona pellucida is missing in the late blastocyst, the 6th day after fertilization.

2.8. Conclusion

During cleavage, the rat ovum undergoes morphologic changes, both qualitative and quantitative, reflected in the structure of the nucleus as well as in that of the cytoplasm. A detailed description of the submicroscopic structure of the one-cell fertilized ovum, of the two-, four-, and eight-cell ova, the morula, and the early and late blastocysts has been given based on material studied at our department.

From the observations of the nucleus, it follows that, besides certain temporary and minor changes concerning the structure of the nuclear envelope, chromatin and/or further inconstant nuclear components, the submicroscopic structure of the nucleoli changes most. In the cytoplasm, more conspicuous and numerous changes of cell or-

---

Fig. 60. Late rat blastoscyst. Trophoblast cell with Golgi complex (G), ribosomes (R), granular endoplasmic reticulum ($\rightarrow$), autophagic vacuole (AV), and remnants of lamellar structures (L). Fixation glutaraldehyde and OsO_4; embedding medium Durcupan ACM, magnification X 24,000

Fig. 61. Late rat blastocyst. Contact of two trophoblast cells with junctional complex ($\rightarrow$) and numerous microvilli and cytoplasmic projections. Fixation glutaraldehyde and OsO_4; embedding medium Durcupan ACM; magnification X 23,000

ganelles and cytoplasmic inclusions can be found. Significant changes take place especially in the structure of mitochondria, the Golgi complex, and the cell membrane, including the formation of cell contacts and junctional structures. Important qualitative and quantitative changes were observed during the differentiation of the rat ovum in the smooth as well as in the granular endoplasmic reticulum, ribosomes, multivesicular bodies, autophagic vacuoles, lamellar structures, and glycogen. The structure of the zona pellucida also changes until its final disappearance. All morphologic transformations occurring during the cleavage of rat ova are an expression of the continuing differentiation and specialization of blastomeres and of the preparation of the whole ovum for implantation.

3. Morphologic and Functional Relationship During the Cleavage of Mammalian Ova

The development of the mammalian ovum in the course of early embryogenesis is characterized not only by morphologic changes described in the preceding section but also by biochemical and functional changes. In this section morphologie and functional relationships will be discussed, based not only on our own results but also on information obtained from the investigation of the ultrastructure (occasionally of the microscopic structure) in ova of various animal species and on comparison of analogous biochemical and experimental data. The information was mostly obtained in lower species of vertebrates and invertebrates, and their applicability to mammals is limited. By summing up current findings concerning the problems of cell surfaces and intercellular contacts, of the nucleus, nucleoprotein structures, and proteosynthesis, of mitochondria and energetic metabolism, of lysosomes, of the occurrence of stored materials and their metabolism in the segmenting ova, it is our goal to further reserarch objectives which should aim at the practical application of the results obtained in veterinary medicine and which should become an impetus for purposeful studies in human medicine.

3.1. Surface Coats and Cell Contacts of the Ovum, Their Differentiation and Function During Cleavage

Milan Dvořák

Eucaryotic cells are separated from the environment by a surface coat which maintains the internal milieu of the cell and plays a vital role in intercellular communication. Two components are recognized as participating in the structure and function of the cell surface (Bennett, 1963). The most basic of the two is the cell membrane, which had long been considered to be the only limiting layer, viz., up to the time of the discovery of the glycocalyx, a structure lying outside the cell membrane and being in close connection with it. In mammalian ova, another component is formed around the cell in the course of oogenesis, the zona pellucida. Both the glycocalyx and the zona pellucida may be called extraneous coats (Stoeckenius and Engelman, 1969). Cell surfaces are differentiated under active participation of the cell membrane, depending on the function pertinent for the specialization of the cell. In multicellular organisms, their representatives are mostly cell contacts and specialized junctions.

3.1.1. Some Remarks on the Structure and Function of the Cell Membrane, the Glycocalyx, and the Zona Pellucida

The basic structure of the cell membrane corresponds to the other biologic membranes of the animal cell (Robertson, 1959, 1960). Its structural basis is a lipid bilayer (Danielli and Davson, 1935), causing the cell membrane to have the appearance of a dark line at a small magnification in the electron microscope and the appearance of a three-layer structure at a great magnification and a good resolution (Zetterqvist, 1956; Sjöstrand, 1956). Recently, new information has revealed the molecular structure of the cell membrane, pointing out its fluid mosaic-like structure, since asymmetrically specific proteins, most of which contain a saccharide component bound to the end of the molecule turned away from the cytoplasm (Capaldi, 1974; Bretscher and Raff, 1975) are included in the lipid matrix. Proteins are responsible for most of the membrane function properties.

The glycocalyx (Bennett, 1963) covers the outer surface of the cell membrane as a finely filamentous layer; its thickness varies from 7.5 - 200 nm (Rhodin, 1974). It is mostly formed by polysaccharides of the mucopolysaccharide type, and to render it visible, a method of staining with ruthenium red was devised by Luft (1971). The general structure and function of the glycocalyx was studied by Ito (1969), Martínez-Palomo (1970), and Rambourg (1971).

The main function of the cell membrane is the preservation of the integrity of the cytoplasm. Further functions of the cell membrane are the transport of small molecules, endocytosis (pinocytosis and phagocytosis), exocytosis, reception and transport of various extracellular chemical signals (such as hormones, growth factors, neurotransmitters, antigens), cell adhesion, and formation of cell junctions. The selective, regulatory, and protective properties of the cell surface are not determined by the cell membrane alone but depend on the combination of cell membrane and glycocalyx (Bennett, 1969; Martínez-Palomo, 1970).

The cell membrane and cell surface can be highly specialized and modified, depending on functional demands and relationship to neighboring cells. If two or more cells in the organism form a more or less constant contact, their surfaces gradually specialize so as to create conditions for maintaining relatively constant mutual relations. Especially cell junctions are formed, represented mostly by a tight junction (the zonula occludens), with its incomplete form of punctate tight junction, gap junction (nexus), intermediate junction (zonula adherens), and desmosome (macula adherens). The classification of these structures is based on the definitions given in the paper by Farquhar and Palade (1963). A peculiar form of desmosomes is a septate desmosome described in the sea urchin blastomeres (Dan, 1960). Cell junctions assert themselves according to type as sites of firm intercellular adhesion; they participate in cellular interactions and are also preferential sites for the crossing of molecules transported from one cell to another.

The zona pellucida as one of the envelopes of the ovum is formed in the course of oogenesis (Chiquoine, 1960). It had previously been hypothesized that the zona pellucida develops from the material of the follicular cells (Sotelo and Porter, 1959; Trujillo-Cenoz and Sotelo, 1959; Björkman, 1962) or that both follicular cells and oocytes contribute to zona pellucida formation (Wartenberg, 1962; Hope, 1965; Norrevang, 1968). Recently, on the basis of autoradiographic analysis, it was proved by Haddad and Nagai (1977) that the zona pellucida is formed by the secretion of material only from the oocyte. The zona pellucida is primarily formed by polysaccharides (Adams and Hertig, 1964). Its structure in the mouse was dealt with in detail by Baraňska et al., (1975). During cleavage, changes in its polysaccharide and protein structure occur, which were found in human ova by Stegner and Wartenberg (1961) and in the ova of the rabbit, rat, and hamster by Chang and Hunt (1956). At the same time, structural changes in the zona pellucida also occur which disappear before implantation.

The zona pellucida is of great importance during oogenesis as a mediator in metabolic processes between the oocyte and follicular cells. After ovulation, the zona pellucida functions as a border between the perivitelline space and the millieu of the tuba uterina (Starck, 1975). The zona pellucida contains species-specific sperm receptor sites (Hartmann and Gwatkin, 1971). A sperm binding to the mammalian zona pellucida represents the mature, fertilizable ovum (Inoue and Wolf, 1975). The block to polyspermy in mammalian fertilization operates both at the zona pellucida and at the ovum cell membrane, although significant species variation exists concerning which of these two sites is more important (Gwatkin, 1976).

3.1.2. Morphology of Ovum Surface During Cleavage

The surface of the ovum is not smooth; the cell membrane projects, both in one-cell and multicellular ova, into microvilli of different lengths. This was found in all mammalian ova. The microvilli are distributed evenly along the whole free surface of the ovum. An exception has been found in the unfertilized mouse ovum, where clear mosaicism of the surface ultrastructure was observed by both scanning and transmission electron microscopy (Eager et al., 1976). A small polar region devoid of microvilli was consistently observed and the structural mosaicism seen in the unfertilized ova was preserved following fertilization. So far, exact quantitative studies of the occurrence of microvilli have not been carried out, but their number decreases during cleavage, particularly at the stage of the late blastocyst. The gradual decrease in the number of microvilli during the cleavage of the mouse ovum was pointed out by Calarco and Epstein (1973), based on a study using the scanning electron microscope. Our observations do not, however, confirm the findings by Schlafke and Enders (1967), viz., that the surface is provided only occasional microvilli as early as in the four-cell rat ovum. This finding is valid for the surfaces of adjacent blastomeres, where in the two-cell ovum the microvilli are unusually numerous but rarely occur in the four-cell rat ovum (Dvořák, 1977).

A constant part of the surface of all stages of development of the rat ovum is the glycocalyx (Dvořák, 1977). The glycocalyx also passes continuously to the surface of cells in intercellular spaces in the two- to eight-cell ovum. The presence of deposits of ruthenium red rendering the glycocalyx visible among trophoblast cells can be proved in only some cases of the early blastocyst. In these ova, the glycocalyx was shown even among embryoblast cells. In the late blastocyst, the glycocalyx was observed only on

the outer surface of trophoblast cells. This difference between the early and late blastocysts apparently depends on the degree of differentiation of the zonulae occludentes
near the apical surfaces of trophoblast cells and their premeability for the stain (Dvořák,
1977). Occasionally ruthenium red was even found among the cells of the late blastocyst. The explanation is offered by Enders and Schlafke (1974), who also found material stained with ruthenium red among trophoblast cells and in the lamina basalis
in the mouse. They point out, however, the fact that in staining with ruthenium red
extensive damage of trophoblast cells often occurs, which can lead to false results.

The surface coat generally shows a smaller affinity for deposition of ruthenium red
in connection with the differentiation of the ovum. The deposits of ruthenium red in
the perivitelline and intercellular spaces are largest and most often occur in the two-
cell stage; they are only seldom present in the eight-cell ovum (Dvořák, 1977).

3.1.3. Development of Contacts and Junctional Structures Between the Cells of the Segmenting Ovum and the Blastocyst

The development of junctional structures has only recently been studied by Calarco
and Epstein (1973) and Nadijcka and Hillman (1974) in the mouse, Hastings II and
Enders (1974b) in the rabbit, Ducibella and Anderson (1975) in the mouse and in the
rabbit, and Dvořák (1978) in the rat ovum. During cleavage, intercellular contacts of
the rat ovum are subject to gradual developmental changes. Morphologically, as the
loosest junction, contact between the blastomeres of the two-cell ovum appears. The
adhesion of cells is increased only by the presence of numerous microvilli distributed
in such a way that they form interdigitations. Amorphous material, completely filling
the intercellular space as proved by staining with ruthenium red (Dvořák, 1977), can
also play an important part in adhesion.

In the four-cell stage, the adjacent surfaces of the cells are almost closely attached
to one another nearly throughout. The first signs of formation of junctional structures
were observed in this stage, which Hastings II and Enders (1974b) denote as rudimen-
tary desmosomes in the rabbit. With respect to the ultrastructural definition of desmosomes (Farquhar and Palade 1963), we do not consider this description adequate. The
penetration of substances between the cells is not limited at this stage, as was shown
by experiments with ruthenium red (Dvořák, 1977) and peroxidases (Dvořák and
Trávník, 1975, 1976).

In the eight-cell stage, the junctional structures are already present. In the vicinity
of the surface of the ovum, they are the close junction and/or the focal tight junction
and the zonula adherens. The penetration of substances between the cells is, however,
not limited even at this stage. In the rat, Enders (1971) considers the first junctional
structure to be the accumulation of dense material in regions of attached membranes;
Schlafke and Enders (1967) call them as prodesmosomes. Unlike these authors, we
found the occurrence of even focal tight junctions, which corresponds to the observations by Ducibella and Anderson (1975) in the mouse morula that the presence of an
apical junction appears to be the fusion of the outer leaflets of membranes. The fact is
that both in the rat and in the mouse individual differences due to the degree of development within a certain stage evidently can exist. Calarco and Brown (1969) did not,
however, notice any specialized structures in the mouse morula. The same reinforce-
ment of the junction among the cells by long cytoplasmic evaginations was observed in
the rat as Calarco and Epstein (1973) observed in the mouse.

The junctions between the cells are most differentiated at the stage of the blasto-cyst. We found the same differences in the junctional structures in the early and late stages of the blastocyst of the rat as Hesseldahl (1971) in the rabbit and Nadijcka and Hillman (1974) in the mouse. There are incomplete zonulae occludentes between tro-phoblast cells in the early rat blastocyst, enabling partial permeation of ruthenium red among the cells of the blastocyst. In the late blastocyst, the occlusion is complete. An-other junctional structure which, however, does not occur constantly is the zonula ad-herens. Desmosomes, such as defined by Farquhar and Palade (1963), were not found in our material. Only in one single case was there a desmosome among trophoblast cells, in which, however, the central plate either had not been formed or was not visible. This is in contrast to the observations by Schlafke and Enders (1967) in the rat and by some other authors in different animals species (Calarco and Epstein, 1973; Hesseldahl, 1971; Hastings II and Enders, 1974b; Nadijcka and Hillman, 1974; Duci-bella et al., 1975), who mention the occurrence of "typical" or "atypical" desmoso-mes. Different data can sometimes be explained by the inconsistency in the exact de-termination of the type of the junctional structure. It is possible, however, that the differences in observations are due to differences in ages (in hours!) of the embryos described. Nadijcka and Hillman (1974) pointed out the quick development of junc-tional structures in the mouse blastocyst within less than 2 days.

Among both trophoblast and embryoblast cells, we found structures corresponding to the zonula adherens and the close junction. Our observation is, to a certain extent, in accordance with the findings of Ducibella et al. (1975) in the ova of the mouse and those of the rabbit. It is, however, necessary to note that Hesseldahl (1971), on the other hand, did not observe any junctional structures in rabbit ova.

We did not observe any characteristic differentiated junctional structures among embryoblast cells. This is in agreement with observations by Schlafke and Enders (1967). Desmosomes which, in other species, are mentioned by Hesseldahl (1971), Hastings II and Enders (1974b), and Ducibella et al. (1975) are not differentiated in the embryoblast of the rat blastocyst.

From our results it follows that from the eight-cell stage onward the adhesion of cells is increased by the presence of junctional structures. We have, however, observed that the cells are firmly attached to each other in some places even in the four-cell stage and in the inner cells of the eight-cell ovum and blastocyst. This phenomenon be-comes particularly conspicuous in artificial dilatation of the intercellular space due to inadequate osmotic conditions during fixation. It is evident that, even where junctio-nal structures are not yet differentiated, local adaptations of the cellular surface come into play which increase the adhesion of cells.

3.1.4. Structure of the Zona Pellucida During Cleavage

The zona pellucida in the rat ova shows morphologic characteristics similar to those described by Barańska et al. (1975) in mouse embryos. For a detailed study of their structure, we have stained with ruthenium red (Dvořák, 1977) and confirmed that the zona pellucida consists of fibrilar and granular material. Four layers of different elec-tron density could be distinguished; in general, the outer part of the zona pellucida has a looser structure than the inner one. During cleavage, the zona pellucida diminishes in thickness, especially as the blastocyst expands (Enders, 1971). Enders points out that, unlike in most other species, the diminution is irregular; it may be quite thin over projecting portions of trophoblast cells and thicker over other regions. Enders and Schlafke (1967) observed − on a submicroscopic level − defects in the zona pellucida through which projections of trophoblast cells penetrated. The content of acid polysaccharides decreases (Dvořák, 1977) simultaneously with the diminution of the thickness of the zona pellucida. This is in agreement with the findings that during the cleavage of the ovum, particularly in the period unfertilized-fertilized ovum (Gordon et al., 1975) and in the blastocyst (Enders and Schlafke, 1974), changes in the composition of the glycoprotein complex of the zona pellucida occur (but also in the surface coat of blastomeres). At the stage of the late blastocyst, in the rat on the 6th (5th) day after fertilization, the zona pellucida disappears. The zona pellucida is surprisingly persistent in blastocysts of the fissiped carnivores (Enders, 1971). Two mechanisms cause its loss (McLaren, 1970): an active hatching mechanism (Dickmann and Noyes, 1961) and lysis (Defrise, 1933; Dickson, 1963; Reinius, 1967). Depending on the circumstances of ovum and uterus, loss of the zona pellucida may involve either one or both of these processes.

From the functional point of view, the zona pellucida forms a certain barrier against the influence of the environment in which the ovum is located (Starck, 1975). As is discussed below, this function of the zona pellucida is selective with respect to the chemical nature of the substance but not to its molecular weight.

3.1.5. Exchange of Substances Between the Segmenting Ovum and the Environment of the Tuba Uterina and the Uterus

The mammalian ovum develops freely in the period of cleavage in the cavities of the tuba uterina and the uterus. The normal process of cleavage can take place, as has been particularly proved in the mouse (Hsu, 1973 and Hsu et al., 1974) and in the rabbit (Brinster, 1970), but in also other mammals in vitro, if suitable conditions of culture are created. The degree of importance of the inner milieu in the tuba uterina and the uterus for the development of the earliest stages of embryos is not exactly known, i.e., only some factors involved here are known. Problems have been gradually studied morphologically as to what substances can penetrate into free segmenting ova and in what way.

Austin and Lovelock (1958), in a pioneering series of experiments, determined the permeability in vitro of the surface coat of rat, hamster, and rabbit ova to toluidine blue, digitonin, alcian blue, and heparin. Since heparin, with a molecular weight of 16,000, could be demonstrated within any of these ova and alcian blue, with a molecular weight of 1200, penetrated into all of them, these authors concluded that the surface of ova consituted a barrier to molecules with a molecular weight greater than 16,000. Glass (1963), however, has demonstrated by use of immunofluorescent methods that serum proteins can pass into the ovum in vivo. The occurrence of proteins present in the lumen of the uterus was also proved biochemically in preimplantation ova (Beier, 1970; Hamana and Hafez, 1970). Enders (1971) studied the ingestion of ferritin, whose molecular weight varies with the iron content between 480,000 and 1,000,000, and of thorotrast, which is a colloidal suspension of thorium dioxide in dextran, forming aggregate particles of undetermined size but appreciably larger than ferritin. He found that both substances permeated the zona pellucida and cell coats of rat ova.

Detailed information on the morphologic picture of ingestion of substances with protein character has been obtained by experiments using horseradish peroxidase and ferritin in the ova of rat, mouse, rabbit, ferret, and guinea pig (Hastings II et al., 1972), peroxidase, myoglobin, and ferritin in the rabbit (Hastings 1973), horseradish peroxidase and ferritin in rats (Schlafke and Enders, 1973), horseradish peroxidase in the rabbit (Hastings II and Enders, 1974a), and peroxidase, ferritin, and myoglobin in the rabbit (Hastings II and Enders, 1974a). Experiments on the ingestion of horseradish peroxidase and microperoxidase by rat ova have also been carried out in our laboratory (Dvořák and Trávník, 1975, 1976).

It appears that there are no basic differences in the ingestion of exogenous proteins in vitro between the individual stages of the ovum, with the exception of the blastocyst. Up to the eight-cell stage, the location of the reaction product of exogenous peroxidases is for the most part in small, micropinocytotic vesicles which only coalesce into vacuoles in isolated cases in later cleavage stages. In the cells of the blastocyst, the reaction product also occurs frequently in lysosomes (Dvořák and Trávník, 1975, 1976). Similarly, it has been observed by Hastings (1973) that the blastomeres of early cleavage stages up to the morula show little uptake of exogenous peroxidase. The occurrence of these enzymes in lysosomes has its beginning as early as the eight-cell stage, which is in accordance both with our findings and with those of Hastings II and Enders (1974a). In the blastocyst, exogenous proteins have been demonstrated both in trophoblast and embryoblast cells. The observations on the occurence of reaction products among trophoblast cells are ambiguous. Occasionally, however, we have found both horseradish peroxidase and microperoxidase among trophoblast cells in our material in the rat, whereas Hastings II and Enders (1974a) have not found horseradish peroxidase in the rabbit blastocyst at this location. Exogenous peroxidase in the blastocyst also passes from the cells into the blastocoel cavity (Hastings, 1973; Dvořák and Trávník, 1975, 1976). The tracer within the cavity is probably a result of transport by trophoblast cells (Hastings, 1973).

From our findings with horseradish peroxidase with a molecular weight of 40,000 and microperoxidase with a molecular weight of 1900, it follows that the ingestion of protein of lower molecular weight is smaller. Endocytosis of exogenous protein is not only dependent on the size of the substance molecule but also on its chemical nature. The selective ingestion of substances of different chemical nature is evidenced by the differences between the endocytosis of horseradish peroxidase and ferritin (Hastings II et al., 1972) and other nonprotein substances (Ferm, 1971). A barrier is formed here both by the zona pellucida, which has been proved, e. g., in the case of heparin (Austin and Lovelock, 1958), and the cell membrane (Hastings II and Enders, 1974a). The zona pellucida also has a specific role; Heyner et al. (1969) found that anti-C3H serum did not cause degeneration of blastocysts of C3H mice if the zona pellucida was intact, but degeneration occurred rapidly in ova lacking the zona. Of importance is the fact that the ovum can be penetrated by large particles such as the viruses (Gwatkin, 1967). The zona pellucida does not represent a substantial obstacle to the penetration of large molecules into the segmenting ovum, as long as their physical and chemical properties do not prevent it.

The free ovum does not only accept substances which can assert themselves favorably in the ovum metabolism but also harmful substances. Adams et al. (1961) have confirmed the findings of the permeability of the rabbit blastocyst to trypan blue as well as its teratogenic effect on the rabbit embryo. Keberle et al. (1965) reported that

^{14}C-thalidomide was able to penetrate into the blastocyst of the preimplantation, implanting, and postimplantation blastocyst.

The basic mechanism of the endocytosis of substances into free ova is pinocytosis (Enders, 1971). Hastings (1973) states that the tracer is ingested by micropinocytotic vesicles and tubules. Gwatkin (1966) has found that Mengo virus particles can enter the mouse ova both with and without the zona pellucida. Since these particles normally have to be phagocytized by the host cell to penetrate it, the results of Gwatkin (1966) are a significant indication of early phagocytic activity. The results of Glass and McClure (1965) with serum protein also illustrate the necessity of a mechanism for engulfing large molecules.

The segmenting ovum is not an isolated object from the milieu of the tuba uterina and uterus. This fact can be of importance for the differentiation processes and metabolism of the ovum. In vitro experiments have demonstrated that serum proteins have a stimulatory effect on trophoblast outgrowth in the mouse (Gwatkin, 1966), increase blastocyst metabolism as judged by carbon dioxide output in this species (Menke and McLaren, 1970), and promote blastocyst growth in the rabbit (Krishnan and Daniel, 1967; Maurer et al., 1970; Gulyas and Krishnan, 1971; El-Banna and Daniel, 1972) and rat (Daniel and Krishnan, 1969). It cannot, however, be excluded that harmful substances can pass into the ovum in the same way, which may even cause pathologic changes in such an early embryo.

3.1.6. Conclusion

The segmenting mammalian ova are separated from the environment of the tuba uterina and uterus by surface coats represented by the zona pellucida and cell membrane with glycocalyx. During the cleavage of the rat ovum, the cell membrane is subjected to changes that are particularly connected to the development of contacts and junctional structures between the blastomeres of segmenting ovum and the blastocyst. Blastomere surfaces facing the perivitelline space are characterized only by a reduction in the number of microvilli during cleavage. The adhesion of blastomeres increases from the eight-cell stage onward caused by the presence of junctional structures which are well-differentiated between trophoblast cells in the late blastocyst. The formation of zonulae occludentes at the stage of the morula is the very condition for the origin of the blastocyst cavity and thus also of the blastocyst. In two-cell and four-cell ova, it is evident that the glycocalyx and possibly also material of similar character play an important role in the adhesion of cells. The glycocalyx is present on the free surface of the blastomeres as well as on the surfaces of the blastomeres in contact. The zona pellucida changes its chemical composition rather than its structure in the course of cleavage and disappears at the late blastocyst stage.

Segmenting mammalian ova are not isolated objects independent of the environment. In passing through the tuba uterina and uterus before implantation, they can accept materials of various chemical characters and even of large molecules, as has been convincingly shown by in vitro experiments. In the control of these processes, both the zona pellucida and especially the cell membrane with the glycocalyx assert themselves. This circumstance can be of importance for the differentiation and metabolic processes of the ovum. The penetration of harmful substances from the outer environment into the free segmenting ovum, particularly from the teratologic point of view, can be a negative characteristic.

3.2. Nucleus, Nucleoprotein Structures, and Proteosynthesis in the Segmenting Ova

Pavel Trávník

The principle of cytodifferentiation is evidently the differential activation of the genes (Davidson, 1969). This process evidently shows the relationships between the nucleus and the cytoplasm whose importance is characterized by Hamburgh (1971) who states that the idea of differential gene activation or repression during development has been practically elevated to the status of a central doctrine of developmental biology. It is thus evident that the key role in differentiation is played by the synthesis of specific proteins whose structure is coded into DNA of nuclei and/or into DNA of further cell organelles. If we want to contribute to the clarification of the processes of differentiation from the morphoplogic point of view, we must concentrate our attention on structures containing nucleic acids. They include the nucleus, ribosomes, granular endoplasmic reticulum, and from the point of view of DNA content, also mitochondria. Individual structural and functional components of the interphase nucleus are the nuclear envelope, the nucleolus, and chromatin. The description of the structure and the functions of the nuclear components have been dealt with recently in a summary paper by Wischnitzer (1973). The morphology of mammalian chromosomes has been synthetically dealt with by Stubblefield (1973). A nuclear component to which maximum attention has been paid in the literature is the nucleolus. In recent years, the monograph by Busch and Smetana (1970) dealt particularly with the problems of the nucleolus, as well as papers by Ghosh (1976) and Sidebottom and Deák (1976). An outline of the information concerning ribosomes is given by De Man and Noorduyn (1969) and Nanninga (1973) and concerning the structure of endoplasmic reticulum by DuPraw (1969). The occurrence, forms, and function of mitochondrial DNA were recently discussed by Korb (1976).

3.2.1. Ultrastructure of the Nucleus, Ribosomes, and Granular Endoplasmic Reticulum in the Segmenting Mammalian Ova

3.2.1.1. Nucleus

No detailed study of the nucleus in mammalian ova has been made as yet since more attention was paid to the other cell organelles. Until now, papers have been chiefly aimed at the structure of the pronuclei in the one-cell ovum and in the other stage at the structure of the nucleolus. The structure of the nucleus in the segmenting rat ovum on the submicroscopic level has been described by Sotelo and Porter (1959), Mazanec and Dvořák (1963), Szollosi (1965a), Schuchner (1970), Szollosi (1971), and Dvořák (1971, 1974a). An outline of light-microscopic findings is given by Austin (1961).

In the fertilized one-cell ovum, two pronuclei are present up to the first cleavage division. The formation of the nucleus of the zygote from a male and a female pronuclei is not know in mammals, perhaps with the exception of the order of monotremata (Austin, 1961). A detailed description of the development of rat pronuclei is given by Austin (1951, 1952, 1961). Pronuclei differ in their size. In the rat, the male pronucleus is larger, as found by Austin (1952) by direct observation of its origin from the head of the sperm. The volume of the male pronucleus is about 2.5 times higher than the female pronucleus; approximately the same relation holds true for the total number and volume of the nucleoli (Austin, 1961). The volume of the pronuclei changes during the development of the one-cell ovum, as well as the number of nucleoli (Sotelo and Porter, 1959). The volume of pronuclei and the number of nucleoli reach their maximum about the middle of the existence of pronuclei (Austin, 1961).

Nuclear Envelope

The pronuclei of the rat ovum are limited by two membranes which form the perinuclear space, 15 - 20 nm wide. The perinuclear space is in some places extremely dilated due to numerous evaginations of the outer nuclear membrane. These dilatations can be as large as 0.5 μm. In the perinuclear space, material of low density is present as well as dark bodies of variable size limited by a simple membrane, in some cases connected with the inner membrane of the nuclear envelope. Szollosi (1965a) also states that in rat pronuclei wide perinuclear spaces are present. Mazanec and Dvořák (1963) state that after fertilization of the rat ovum invaginations of the nuclear envelope originate.

During the cleavage of the rat ovum, numerous changes occur in the morphology and cytochemistry of the nuclear envelope. At the two-cell stage, numerous invaginations and evaginations are still present on the nuclear envelope; at the four-cell stage, only local dilatations of nuclear envelope remain. This finding is in accordance with Mazanec and Dvořák (1963). In later stages, the width of the perinuclear space changes only little. Nuclear pores are irregularly distributed in all stages of the rat ovum and are filled with electron dense material. In the eight-cell ovum, the first ribosomes appear on the nuclear envelope. This finding is in accordance with Schlafke and Enders (1967). The number of ribosomes on the nuclear envelope increases in later stages. At the eight-cell stage, signs of the blebbing processes also appear, "blebs" also being found in later stages.

By electron cytochemical methods, the activities of acid phosphatase (Šťastná, 1977a), alkaline phosphatase (Šťastná, 1979), and organophosphate-sensitive esterase (Trávník, 1977b, 1978) have been found in the perinuclear space of the segmenting rat ovum. In the one-cell ovum, all the enzymes mentioned above occur. This state lasts up to the eight-cell stage. In the early blastocyst, alkaline phosphatase activity disappears as well as most of the activity of organophosphate-sensitive esterase, which is present in only a part of trophoblast cells. Acid phosphatase activity can be established throughout the cleavage.

Chromatin

Chromatin of the rat pronuclei is regularly distributed, the nucleus appears light, and condensations are visible only in places which, however, do not exhibit the character of karyosomes. A similar picture is found by other authors in other animal species. In human pronuclei, chromatin makes an impression of great hydration, and in places it is condensed into irregular bundles of dense filaments, most of which are located near the pronuclear envelope (Zamboni et al., 1966a, b). In the rabbit, pronuclear chromatin is less dense and has the form of dispersed granular material. The condensed zones of chromatin are visible at the pronuclear envelope and rarely near the nucleolus (Zamboni and Mastroianni, 1966b). The distribution of chromatin in the nucleus of the pig ovum is asymetric, and the so-called "nucleospheridies" are present in the nucleus (Norberg, 1973a, b). Hillman and Tasca (1969) claim that the nucleus of the mouse ovum has essentially a homogenous appearance and contains granules similar to inter-chromatin granules. Szollosi (1971) states that interchromatin as well as perichromatin granules are present.

In the course of the cleavage of the rat ovum, chromatin remains nearly regularly distributed up to the stage of the early blastocyst, but the number of cluster-like aggregations of chromatin increases. Typical small karyosomes appear in the late blastocyst both in trophoblast and embryoblast cells.

In the eight-cell ovum of the mouse, aggregations of chromatin appear, often associated with the nucleolus and the inner membrane of the nuclear envelope and/or intranuclear annulate lamellae (Hillman and Tasca, 1969). In the mouse blastocyst, condensed chromatin is usually associated with the inner membrane of the nuclear envelope (Nadijcka and Hillman, 1974). In the rabbit, chromatin is dispersed in the nuclei of the blastocyst (Hesseldahl, 1971).

Nucleolus

At the stage of the pronuclei, a great number of nucleoli were found in the rat ova. In an ultrathin section of one pronucleus, as many as ten nucleoli were found. Austin (1952) ascertained a maximum of 36 nuclei in one pronucleus and Sotelo and Porter (1959) observed 12 - 17 and more nucleoli. Relatively frequently, one big nucleolus with a diameter of 6 μm is present in the pronucleus. In addition, more nucleoli with a size of about 2 μm and a high number of nucleoli of substantially smaller diameter are present in the pronucleus. All nucleoli at this stage are compact, formed by the fibrillar component. Dvořák (1974a) sometimes observed nucleoli with an indication of granular component formation. The nucleoli are located freely in karyoplasm, partly attached to the inner membrane of the nuclear envelope. The ultrastructure of the nucleolus in the rat ovum, particularly in the pronuclear phase, was dealt with in detail by Szollosi (1965a) and Schuchner (1970). Schuchner (1970) states that before the rise of typical, extremely dispersed chromatin of the pronuclei a large central nucleolus is formed; small bodies not touching the nuclear envelope were also sometimes observed. They are identical with primary nucleoli, as described by Austin (1961). Later on, nucleoli attached to the nuclear membrane were observed which were associated with the inner membrane of the nuclear envelope by a large portion of their surface. Some of them were observed in the protrusions of the nuclear envelope. Those nucleoli are identical with the secondary nucleoli according to Austin (1961). Much smaller nucleoli attached to the nuclear membrane are identical with tertiary nucleoli according to Szollosi (1965a). In the later period of the existence of the pronuclear stage, the number of secondary nucleoli increases, and then the number of large nucleoli attached to the nuclear envelope also increases. Extrusions of them have been found. Finally, 24 h after fertilization, one large central nucleolus can frequently be observed. Schuchner (1970) further states that the nucleolus at this period is formed by thin coiled fibrils with a diameter of 10 nm which emerge from the periphery of the nucleolus. Inside they are more closely packed, forming a homogenous structure. Structures similar to typical granules from somatic cells are not distinct. In all types of nucleoli, RNA was proved (Szollosi, 1965a, 1971) and around the nucleoli DNA was ascertained (Alfert, 1950; Austin, 1961; Szollosi, 1965a). Great attention was paid to the extrusion of nucleolar material in rat pronuclei. It is described by Austin (1961), Szollosi (1965a), and Schuchner (1970).

In the mouse ovum in the pronuclei, spheric nucleoli without the pars granulosa with a small amount of condensed chromatin in their surroundings were observed. Tertiary nucleoli were not found (Calarco and Brown, 1969). In the hamster ovum, 1 - 17 nucleoli were observed in the pronuclei and later 1 - 6 nucleoli (Pickworth et al., 1968). In rabbit pronuclei, numerous nucleoli were present near the membrane and aggregations of chromatin were seldom observed near the nucleoli (Zamboni and Mastroianni, 1966b). In rabbit pronuclei, the process of nuclear extrusion was also observed (Gulyas, 1971). In pig pronuclei, several filamentous, compact nucleoli were found (Szollosi and Hunter, 1973). In the pronuclei of *Peromyscus maniculatus* only one nucleolus usually occurs, in the pronuclei of *Meriones unquiculatus* one to three nucleoli (Marston and Chang, 1966). In human pronuclei, one compact nucleolus has been established (Zamboni et al., 1966a, b).

According to our observations, changes occur in the number and morphology of nucleoli during cleavage. In the two-cell stage, the number of nucleoli decreases when compared with the one-cell stage; in a section, one to three but most often one to two nucleoli are usually present. This number is constant throughout the whole cleavage up to the stage of the blastoscyst. In the two-cell and four-cell ova, ring-formed nucleoli were sometimes found, a fact which does not quite agree with the conclusions reached by Sotelo and Porter (1959), who consider ring-formed nucleoli typical of the two-cell stage. The changes in the structure of the nucleolus are the gradual appearance of nucleolonema, the increase in pars granulosa, and a change in the nucleoli from compact to reticular. Quantitative evaluation of the individual types of nucleoli in certain stages was carried out by Dvořák (1974a).

Schuchner (1970) states that the morphology of the nucleolus changes in the course of cleavage and nucleoli can no longer be included into the classification of primary, secondary, and tertiary nucleoli. Her observations about the development of nucleolar structure in the course of cleavage agree with ours. Morphologic changes with a gradual appearance of the nucleolonema in the mouse are described by Calarco and Brown (1969) and Maraldi and Monesi (1970). Hillman et al. (1967) divide each stage of the segmenting mouse ovum cultured in vitro into the early, middle, and late stages, stating that the reticulation of the nucleolus always appears in the following stage of the earlier period. Szollosi (1971) also explains the differences in the morphology of the nucleolus of various stages of cleavage of the mouse and rat ova by the fact that reticulation arises around the compact nucleolus in each following stage in the earlier phase of the cell cycle, the period of existence of the reticular nucleolus becoming increasingly longer. He also states that at all stages investigated, from the two-cell ovum up to the morula, usually only one of several nucleoli is morphologically transformed at a time. In his opinion, it cannot be determined from available studies whether every nucleus is subject to transformation during the cell cycle. Gonzáles and Nardone (1968) prove the existence of the defined cycle of morphologic changes in the nucleolus during the cell cycle. Irregular morphology, marked association with the nuclear envelope, and the movement of the nucleolus were characteristic of G_1 and G_2 phases of the cell cycle. During the S phase, the nucleoli are oval, located centrally, and exhibit much less movement. The changes of the nucleolar morphology dependent on the cell cycle are observed even in other types of cells.

3.2.1.2. Ribosomes and Granular Endoplasmic Reticulum

According to our findings, the quantitative representation of ribosomes and the granular endoplasmic reticulum during the cleavage of the rat ovum changes significantly. In the one-cell ovum, only a small quantity of ribosomes is present, and this situation persists up to the four-cell ovum. Polysomes were found in the two-cell ovum only in the vicinity of the mitotic spindle and in the four-cell ovum in isolated cases also elsewhere in the cytoplasm. In the eight-cell ovum, the number of ribosomes evidently increases; they are located mostly in the vicinity of the nucleus and the mitochondria; they have the form of polysomes and are partly attached to the outer membrane of the nuclear envelope and the membrane of the granular endoplasmic reticulum. The number of ribosomes further increases in the morula and the early and the late blastocysts, when the increase in the number of polysomes becomes conspicuous. At the stage of the late blastocyst, the areas of the cytoplasm filled with ribosomes are larger in the embryoblast than in the trophoblast. The structures of granular endoplasmic reticulum were found with certainty at the eight-cell stage for the first time. The number of ribosomes attached to the membrane is still low there; part of the structures of granular endoplasmic reticulum are attached to the mitochondria. In the early blastocyst, granular endoplasmic reticulum is located in zones containing cell organelles, and the number of ribosomes attached to membranes at regular distances increases. In the late blastocyst, the quantity of granular endoplasmic reticulum increases; in trophoblast cells, its amount is greater than in embryoblast cells. These observations essentially agree with the findings made by Schlafke and Enders (1967). In the eight-cell ovum in the granular endoplasmic reticulum organophosphate-sensitive nonspecific esterase, which sometimes even occurs in the early blastocyst (Trávník, 1977b, 1978), further alkaline phosphatase (Šťastná, 1979), and acid phosphatase which persists even at early and late blastocyst stage (Šťastná, 1977a) have been observed.

The presence of ribosomes before and after the fertilization of the mammalian ovum was studied in the mouse. Maraldi and Monesi (1970) state that before the fertilization numerous ribosomes and polysomes and granular endoplasmic reticulum are present; after the fertilization, they are found only rarely. Calarco and Brown (1969) state that both in the unfertilized and in the fertilized mouse ovum groups of ribosomes were observed, but neither granular endoplasmic reticulum nor ribosomes were found near the nuclear membrane. Zamboni and Mastroianni (1966b) found granular endoplasmic reticulum in the rabbit ovum at the pronuclear stage. Zamboni et al. (1966a, b) state that the human ovum at the pronuclear stage only rarely contains ribosomes associated with membranes. Szollosi (1967) claims that in the just fertilized mouse ovum polysomes are present. Ribosomes in relatively rich quantities are also present in the hamster ovulated ovum (Longo, 1974). In the pig ovum, small groups of ribosomes are present at the pronuclear stage (Szollosi and Hunter, 1973).

In the early stages of segmenting ova of mammalian species studied so far, few ribosomes are present; they are mostly not aggregated into polysomes, and granular endoplasmic reticulum is usually not present, as follows from our findings and is evidenced by Calarco and Brown (1969), Hillman and Tasca (1969) and Maraldi and Monesi (1970) in the mouse, Hesseldahl (1971) in the rabbit, Panigel et al. (1975) in the baboon, and others. During further development of the ova, the number of ribosomes gradually increases, and granular endoplasmic reticulum appears. In the mouse, the first cisternae of granular endoplasmic reticulum are described in the four-cell ovum (Calarco and Brown, 1969), in the rat in the eight-cell ovum (Schlafke and Enders, 1967), and in the baboon in the blastocyst (Panigel et al., 1975). At the stage of the morula or the blastocyst, granular endoplasmic reticulum is described in all the animal species studied. Enders (1962) observed it in the armadillo, Calarco and Brown (1969), Maraldi and Monesi (1970), and Nadijcka and Hillman (1974) in the mouse, Schlafke and Enders (1967), Šťastná (1972), and Wu and Meyer (1974) in the rat, Hesseldahl (1971) in the rabbit, and Panigel et al. (1975) in the baboon. At the stage of the eight-cell ovum up to the blastocyst, an increase in the number of ribosomes and polysomes is described, as well as in the amount of granular endoplasmic reticulum (Calarco and Brown, 1969; Hillman and Tasca, 1969; Maraldi and Monesi, 1970; Hesseldahl, 1971; Potts and Racey, 1971; Nadijcka and Hillman, 1974; Karp et al., 1974). Enders (1971) considers the increase in the amount of polysomes and granular endoplasmic reticulum in the blastocyst to be a general phenomenon. Karp et al. (1974) think that the number of ribosomes can be a limiting factor of the proteosynthesis in the ovum. Szollosi (1971) states that the increase in polysomes in the cytoplasm goes hand in hand with the changes in the morphology of the nucleolus.

Morphologic findings confirm the conclusion made by Epstein (1975) on the basis of biochemical findings. He states that the proteosynthesis in mammalian segmenting ova plays an important role throughout cleavage, unlike in other animal species. A significant decrease in the number of ribosomes about the time of ovulation witnesses the reconstruction of the proteosynthetic apparatus of the ovum. When major changes in the biosynthetic function of the cells occur, the ribosomes present are destroyed and new ones are synthesized (Grasso and Woodard, 1966; Cocucci and Sussman, 1970). A significant rise in the number of polysomes and the appearance of granular endoplasmic reticulum in the eight-cell ovum give evidence of a significant increase in proteosynthetic activity at this stage.

3.2.2. Synthesis of DNA, RNA, and Proteins in the Segmenting Mammalian Ova

Much attention has been paid — besides morphologic studies — by a number of authors to the study of the metabolism of proteins and nucleic acids. The dynamics of the synthesis of RNA, DNA, and proteins has been studied and/or their quantitative spectra, particularly when using the incorporation of radioactive precursors. Even inhibitors affecting the synthesis of nucleic acids and proteins at different levels have been used and the influence of hormones on the incorporation of radioactive precursors studied. Almost all experiments of this kind have been made in the mouse and in the rabbit, occasionally in other mammalian species. Thus, the possibility of generalizing the results of the biochemical approach is greatly limited. A survey of biochemical papers published is given Biggers and Stern (1973) and Epstein (1975).

3.2.2.1. DNA Synthesis

DNA synthesis has been mainly studied by the incorporation of ^{3}H-thymidine into the nuclei of the segmenting ovum and then by photometric methods. Mintz (1964) states that mouse ova have the ability to incorporate thymidine into DNA throughout cleavage. In the rabbit ovum, the synthesis of DNA has been proved as early as in the pronuclei, within 3 - 6 following fertilization. The presynthetic and synthetic periods are short, and the postsynthetic period is long (Szollosi, 1966). Manes (1969) states that

^{3}H-thymidine incorporation in the rabbit is proportional to the number of blastomeres at the respective stage. ^{3}H-thymidine incorporation in the rabbit ovum takes place in the vicinity of the intranuclear annulate lamellae, not in the vicinity of cytoplasmic ones. Quantitative studies of DNA content in the individual stages show a certain excess of DNA in the mouse ovum when compared to the diploid amount. Excess DNA is considered extranuclear and cytoplasmic (Olds et al., 1973). Photometrically, it was proved that DNA content was doubled in the interphase of each mitosis of the mouse ovum (Alfert, 1950) and the mouse and rat ovum (Dalcq and Pasteels, 1955). Barlow et al. (1972) state that all nuclei of the five-cell to sixteen-cell mouse ovum contain a diploid amount of DNA, all of which incorporate ^{3}H-thymidine. After a short-time exposure to ^{3}H-thymidine, unlabeled nuclei were also found which, at the time of precusor application, were probably in the G_1 or G_2 phase of the cell cycle. Blastocysts in vivo and cultured in vitro contain nuclei with a more than diploid quantity of DNA; these polyploid nuclei belong to the cells of primary trophoblast.

3.2.2.2. RNA Synthesis

Quantitative data by different authors concerning the content of RNA in the segmenting ova differ significantly (Biggers and Stern, 1973). Several differences between the ova of mammals and those of lower animals are found in RNA synthesis (Golbus et al., 1973). The incorporation of ^{3}H-uridine into the segmenting ovum has been proved by a number of authors and was demonstrated by Woodland and Graham (1969) in the two-cell mouse ovum cultured in a medium containing fetal calf serum. Hillman and Tasca (1969) and Tasca and Hillman (1970) observed the incorporation of ^{3}H-uridine into the nuclei of the two-cell mouse ovum even in synthetic media, while Monesi and Salfi (1967) and Ellem and Gwatkin (1968) did not establish the incorporation in synthetic media. Mintz (1964) states that the incorporation of ^{3}H-uridine into the pronucleus of fertilized ovum is small and that it increases in the two-cell ovum, where particularly the nucleolus is labeled. During cleavage, both the incorporation of uridine and RNA synthesis increase, as confirmed in the papers by Monesi and Salfi (1967), Ellem and Gwatkin (1968), Manes (1969), Graves and Biggers (1970), Pike et al. (1975). The total amount of RNA polymerase in the mouse ovum also increases (Siracusa, 1973).

The individual types of RNA are synthesized to different degrees in various stages of the segmenting process. Knowland and Graham (1972) and Woodland and Graham (1969) state that newly synthesized rRNA in the mouse embryo was first detected at the four-cell stage. Clegg and Pikó (1975) found the synthesis of rRNA as early as the two-cell mouse ovum and state that in the four-cell ovum it forms the majority of RNA synthesized. Pike et al. (1975) state that the highest incorporation of labeled CO_2 in all stages was also found in rRNA. In the rabbit, unlike in the mouse, seven segmenting mitoses pass without rRNA synthesis, the first indications of synthesis being found here in the transformation of the morula into the blastocyst. Another type of RNA whose synthesis has been studied is tRNA. Woodland and Graham (1969) state that newly synthesized tRNA was isolated for the first time in the four-cell mouse ovum, while Clegg and Pikó (1975) found tRNA synthesis as early as the two-cell mouse ovum. Pike et al. (1975) did no find labeled CO_2 incorporation in this type of RNA in the segmenting mouse ovum. In the rabbit, tRNA synthesis was found in all stages of development (Manes, 1971). If the beginnings of the synthesis of tRNA are compared with that of rRNA, then the synthesis of both tRNA and rRNA in the mouse begins in the same cell cycle (Woodland and Graham, 1969); in the rabbit, there is an interval of six cell divisions between the beginning of tRNA and rRNA synthesis. The last type of RNA studied extensively is mRNA. Clegg and Pikó (1975) observed the synthesis of mRNA in the mouse ovum from the two-cell stage, Golbus et al. (1973) state that mRNA in the mouse is probably synthesized from the one-cell stage and is essential for the ripening of the nucleolus, its integrity, and function. In the rabbit, the findings concerning the synthesis of mRNA are rather diverse. While Manes (1971) states that mRNA is synthesized at all preimplantation stages, Schultz et al. (1973) and Schultz (1973) state that the synthesis of mRNA is detectable at the 16-cell stage and that newly formed mRNA is associated with polysomes.

The inhibitors of the synthesis of RNA and of proteins significantly influence the development of the ovum. Golbus et al. (1973) state that actinomycin D is an effective inhibitor of the blastocyst differentiation and mitosis from the one-cell stage of the mouse ovum. Skalko and Morse (1969) state that high doses of actinomycin D inhibit the synthesis of all RNA and prevent the formation of the blastocyst in the mouse. Unger and Dickson (1971) also state that actinomycin D has the ability to block the differentiation of the blastocyst. The influence of actinomycin D results in changes in the nucleolus of the four-cell and eight-cell ova (Mintz, 1964). Further substances also influence the development of mouse embryos: puromycin (Thomson and Biggers, 1966), cycloheximide (Unger and Dickson, 1971), α-amanitine (Golbus et al., 1973), and fluorophenylalanin (Thomson and Biggers, 1966).

3.2.2.3. Proteosynthesis

Like the synthesis of nucleic acids, the synthesis of proteins has also been studied both from the qualitative and quantitative aspects. Quantitative determination of the total amount of proteins in the rat and in the mouse ova was carried out by Schiffner and Spielmann (1976). They found that in the mouse there were two maxima in the content of proteins, in the two-cell ovum, and in the late blastocyst, whereas in the rat there was a continuous decrease in the content of proteins from the one-cell ovum up to the late blastocyst. A number of authors have provided evidence for the fact that proteosynthesis takes place throughout the preimplantation development of the mammalian ova (Mintz, 1964; Monesi and Salfi, 1967; Monesi et al., 1970; Pikó, 1970; Tasca and Hillman, 1970; Weitlauf, 1971).

In the course of the cleavage in mammalian ova, proteosynthesis increases, particularly at the blastocyst stage (Monesi and Salfi, 1967; Weitlauf and Greenwald, 1967; Manes, 1969; Manes and Daniel, 1969; Graves and Biggers, 1970; Biggers and Stern, 1973; Epstein and Smith, 1973; Karp et al., 1974). Qualitative changes in the spectrum of the proteins produced have also been studied. In the mouse, incorporation of amino acids was found during cleavage into 20 - 25 proteins, of which five were quantitatively the most important. In three of them the synthesis increases during development, in one of them it decreases, and in another one it does not change (Epstein and Smith, 1974). Brinster et al. (1975) state that in the mouse there are small differences between the quality of synthesized proteins of the one-cell unfertilized ovum and the fertilized one. Reproducible differences were found between the one-cell fertilized ovum and two-cell ovum. Significant differences were found between proteins synthesized on the two-cell ovum and in the blastocyst. Biggers and Stern (1973) state that specific proteins necessary for development are also synthesized in the one-cell unfertilized ovum or the two-cell ovum. Petzold (1974a, b) also found qualitative and quantitative changes in the proteosynthesis between the individual stages of development of the rabbit blastocyst and between the trophoblast and embryoblast. Tucker and Schultz (1976) found 550 polypeptides synthesized at this stage in the 6-day-old blastocyst of the rabbit. Van Blerkom and Manes (1974) state that most qualitative changes in the proteosynthesis were observed during the cleavage of the rabbit ovum, the period of development of the blastocyst being characterized by a uniform and constant quality of proteosynthesis. The spectrum of proteins in cultured ova corresponds to the spectrum of proteins produced in vivo at the corresponding stages. Karp et al. (1974) state that the synthesis of ribosomic proteins increases during cleavage together with the increase in RNA synthesis.

In many papers, the influence of estrogens and gestagens on the synthesis of nucleic acids and proteins has also been studied. An outline of the papers on the influence of hormones is given by Holmes and Dickson (1975). A number of papers support the fact that particularly at the blastocyst stage the development of the ovum is influenced by hormones (Weitlauf and Greenwald, 1965; Sugawara and Hafez, 1967; Prasad et al., 1968; Smith, 1968; Weitlauf and Greenwald, 1968; Dass et al., 1969; Manes and Daniel, 1969; Prasad et al., 1969; Weitlauf 1969a, b; Jacobson et al., 1970; Sanyal and Meyer, 1970; Inoui, 1971; Mohla and Prasad, 1971; Wu and Meyer, 1974; Holmes and Dickson, 1973). Some findings, however, support the fact that the influence of hormones on the blastocyst is mediated by the endometrium (Weitlauf, 1971), while in other cases, the influence of hormones has not been found (Jones et al., 1976).

3.2.3. Conclusion

During cleavage, a number of conspicuous changes occur, some of which can presently be interpreted from the morphologic and functional points of view while others cannot yet be interpreted. The dynamic development of the structure of the nucleolus, a great number of nucleoli, and indications of nuclear extrusion observed in the earliest stages of the development of the rat ovum testify to considerable activity of the nucleolus at that period, even though its importance is not yet known. Considerable attention will have to be paid to research on the chemical composition and morphology of the nucleolus at this period. Changes in the morphology of the nucleolus in later stages of the segmenting ovum are also conspicuous. They can be connected to the increasing synthesis of rRNA and eventually to a growing number of ribosomes in the course of cleavage of the rat ovum. The morphology of the nucleolus must, however, be evaluated with respect to the fact that it depends on the phase of the cell cycle, and the dynamics of the cell cycle must be taken into consideration in the study of the morphology and cytochemistry of the nucleolus.

An abrupt decrease in the amount of ribosomes during the period of fertilization is considered to be an expression of the reconstruction of the proteosynthetic apparatus. The actual number of ribosomes and polysomes present in the earliest stages of the segmenting ovum have not yet been fully clarified. For a more exact quantitative evaluation of the presence of ribosomes, it will be necessary to use their selective demonstration. The significant increase in the amount of polysomes and the appearance of granular endoplasmic reticulum in the eight-cell ovum provide evidence of a distinct increase in proteosynthetic activity at this stage. Morphologic findings, however, favor the fact that proteosynthesis — at least to a small extent — takes place even in early segmenting stages of the rat ovum, which is in agreement with the results of biochemical papers on other mammalian species.

3.3. Mitochondria and Energetic Metabolism of Segmenting Ova

Drahomír Horký

Living cells are continuously in the process of rebuilding, gradually decaying, and renewing individual molecules as well as whole structures. From the energetic point of view, they are in the state of dynamic equilibrium, for whose maintenance a continual supply of energy is required.

3.3.1. Principles of Energetic Metabolism of the Cell

Energy is obtained from organic compounds by catabolic reactions. Each form of work performed in the cell is a metabolic pathway in which energy is consumed. The problem of utilizing the energy of organic compounds for performing work is a problem of coupling the individual enzyme reactions so that the energy released in one reaction is available for another reaction. A condition for such coordination is the existence of some common intermediary product which is formed in one reaction and utilized in the other. In living organisms, a universal system of energy transfer is represented through the adenylate system consisting of AMP, ADP, and ATP.

ATP serves as an immediate source of energy for various biosynthetic processes as well as physiologic functions requiring the supply of energy, such as muscular work, transport across membranes, conduction of excitation impulses, etc. ATP can be formed either in the direct reaction without oxygen participation in the substrate phosphorylation or, in connection with the respiratory chain, by the oxidative phosphorylation which, from the energetic point of view, is much more effective.

The eukaryotic cells obtain most of the necessary energy by aerobic, i.e., oxidative degradation of foodstuffs connected with the consumption of oxygen and formation of CO_2 and H_2O. Carbon and hydrogen atoms from organic substances are transferred in this process to their energetically lowest, maximally oxidized level. In this process a number of enzymes participate organized in the multienzymatic complexes. The cycle of tricarboxyl acids (the Krebs cycle) and the respiratory chain are the most important factors for energetic metabolism.

The enzymes participating in these two processes are localized inside the mitochondria. The enzymes of the Krebs cycle are dissolved inside the mitochondria in the mitochondrial matrix. The most important are pyridine-linked dehydrogenases which, in the course of the cycle, produce reduced nicotinamide-adenine dinucleotide coenzymes ($NADH_2$). Some of them, i. e., succinate dehydrogenase, are firmly bound to the structure of the inner mitochondrial membrane. The succinate dehydrogenase also participates in the Krebs cycle and produces reduced flavin coenzymes ($FADH_2$). Another product of the Krebs cycle is CO_2.

Reduced coenzymes are reversible oxidized in the respiratory chain. This is an enzyme complex by which the transfer of hydrogen atoms is performed from reduced coenzymes to oxygen. Through the first links of this complex, i. e. dehydrogenases and ubiquinone, hydrogen atoms are transferred; later only electrons pass through the cytochrome system consisting of the b and c types of cytochromes and cytochromoxidase. The final products of this oxidation are water molecules. The accessibility of the individual enzymes of the respiratory chain is ensured by their structural organization in the inner mitochondrial membrane (Heldt, 1972).

The transfer of hydrogen to oxygen is carried out in the direction of the potential gradient. Thus, free energy is obtained which is stored for the needs of the organism and converted into ATP. The oxidation of substrate hydrogen by oxygen to water in the course of the respiratory chain quantitatively represents one of the most important pathways among the metabolic reactions, by means of which aerobic organisms produce chemical energy from foodstuffs.

It is commonly assumed that some intermediates exist in the ATP formation which represent another form of energy. At present, most of the observed facts support the so-called chemiosmotic hypothesis of ATP origin formulated by Mitchell (1961, 1965). According to this theory, this intermediate form of energy could be a gradient of protons, constituted by the separation of charge across the mitochondrial membrane. Another hypothesis, the hypothesis of a macroergic intermediate (Slater, 1953, 1971), assumes the existence of some energetically rich intermediate, i. e., a concrete chemical compound similar to that functioning in the substrate phoyphorylation. The most recent hypothesis of ATP origin was postulated by Boyer (1965), who assumes some conformation changes in the mitochondrial membrane as a course of its energization.

The coupling of the respiratory chain to the oxidative phosphorylation is so tight in intact mitochondria that the respiration does not proceed without simultaneous phosphorylation of ADP (Chance and Williams, 1956; Lehninger, 1964). ATP originating in mitochondria is consumed in the extramitochondrial compartment. The transfer processes of ATP from mitochondria into cytoplasm across the mitochondrial membrane are another, at present widely studied, field of energetic metabolism.

3.3.2. Changes in Mitochondria During the Cleavage of the Mammalian Ovum

Mitochondria of differentiated cells are of a more or less constant appearance, influenced by the changes of cellular metabolism, above all by the metabolism of ADP and ATP, the exchange of ions, water, and in some cases, even hormones (Chappel, 1966; Hackenbrock, 1967, 1968; Harris et al., 1968; Schnaitman and Greenawalt, 1968; Brinster, 1971; Stern et al., 1971; Lehninger, 1972; Biggers and Stern, 1973; Ginsberg and Hillman, 1973). There are also changes in the segmenting mammalian ovum connected with the differentiation of blastomeres, and changes occur both in the shape and number of mitochondria.

If we sum up the data about mitochondria in the segmenting rat ovum given in section 2, we can state that in the early stages of cleavage of the rat ovum, from the one-cell stage up to the four-cell stage, mitochondria of round to oval shape, with an average size of 0.5 - 0.7 μm are present. The matrix is electron dense, mostly containing a small number of mitochondrial cristae which run in an arch-like way, often bridging the mitochondria (Mazanec and Dvořák, 1963; Schlafke and Enders, 1967). During this period, not even their volume density in the ova changes substantially (Dvořák et al., 1977). Mitochondria in these stages have a close relation to the vesicles of smooth endoplasmic reticulum.

The greatest changes in the structure of mitochondria occur in the eigth-cell stage and blastocyst stage. In the mitochondria of the eight-cell stage and blastocyst cristae appear consisting of three membranes (Schlafke and Enders, 1967); sometimes, there is a constriction of mitochondria (Izquierdo and Vial, 1962), while the volume density pro ovum — when compared to the preceding stages — remains the same (Dvořák et al. 1977). At this stage, mitochondria establish contact with granular endoplasmic reticulum.

At the stage of the early blastocyst, most of the mitochondria of embryoblast cells have an oval to considerably elongated shape (Procházka and Mazanec, 1965; Schlafke and Enders, 1967); often, branched or constricted mitochondria also appear. Some of them are as long as 1.5 μm and contain numerous, parallelly oriented cristae, running perpendicular to the longitudinal axis of the mitochondria (Dvořák, 1971; Šťastná, 1972). In some cases, the cristae are formed by three membranes (Schlafke and Enders, 1967). Mitochondria in trophoblast cells are oval with infrequent, irregularly arranged cristae (Šťastná, 1972).

At the stage of the late blastocyst, almost all mitochondria in the embryoblast cells have a filamentous appearance, reaching a length of 2 - 4 μm, while the volume density of mitochondria remains unchanged in comparison to earlier stages (Dvořák et al., 1977). Frequently, constricted or branched mitochondria are present, containing numerous, transversally oriented cristae (Schlafke and Enders, 1967), or in isolated cases, arch-like cristae. Mitochondria are in close contact with the cisternae of granular endoplasmic reticulum (Dvořák, 1971; Šťastná, 1972). In trophoblast cells, mitochondria have the same appearance as those in the early blastocyst (Šťastná, 1972).

Very similar changes in number, shape, and inner arrangement of mitochondria can also be observed in other mammalian species in the same periods of cleavage. In the one-cell to four-cell mouse ovum, mitochondria with the same appearance were described by Calarco and Brown (1969), while Maraldi and Monesi (1970) and Pikó and Matsumoto (1976) found small, dense mitochondria. In addition, the latter authors describe the same vacuolization as Hillman and Tasca (1969) and Pikó and Chase (1973). As for the data concerning the occurrence of mitochondria without cristae (Maraldi and Monesi, 1970), that observation is most probably distorted by the presence of the electron dense matrix which makes the exact differentiation of cristae very difficult in some cases. The vacuolization of the matrix in mouse ova is probably due only to dilatations of mitochondrial cristae which then imitate vacuoles in the matrix and not the formation of vacuoles, as stated by Pikó and Matsumoto (1976). In other mammalian species, such as pigs, the formation of vacuoles in the mitochondria during the one-cell to the four-cell stages of cleavage is a regular phenomenon. It is necessary to understand this phenomenon as species conditioned (Norberg, 1973a, b), the same as structures consisting of double membranes in the mitochondrial matrix of the rabbit ovum (Krauskopf, 1968).

Just as in the rat ovum, there are also changes in the ova of other mammalian species not only in the distribution of mitochondria but also in their appearance at the eight-cell stage, when Hillman and Tasca (1969) found a mixed population of mitochondria, mostly elongated, in the mouse ovum at this stage of cleavage.

At the stage of the blastocyst, mitochondria of embryoblast cells of the rat ovum as well as those of other mammalian species are very similar in appearance, whereas the mitochondria of trophoblast cells are different in shape, according to the observations made by Maraldi and Monesi (1970), Hesseldahl (1971), McReynolds and Hadek (1972a), Aitken et al. (1973), and Nadijcka and Hillman (1974).

93

The described changes in the number, shape, and arrangement of mitochondria are very similar in all mammalian species studied so far, and they can be considered – including the changes in the arrangement of their inner structure – to be a general feature of the early stages of segmenting ova (Hadek, 1969; Enders, 1971; Stern et al., 1971; Szollosi and Hunter, 1973). It is probable that the change in their shape from circular and/or oval forms in the first stages of cleavage elongated to rod-like forms is an indication of differentiation of this organelle in the course of cleavage, while the changes in the number, arrangement, and the course of cristae can be associated with metabolic processes. These relationships can be derived only from the observations of similar changes which take place in the mitochondria of other kinds of cells in the course of their differentiation (Dvořák, 1971). The changes in the number, course, and arrangement of cristae associated with metabolic processes were found in a suspension of mitochondria (Hackenbrock, 1967, 1968; Lehninger, 1972; Biggers and Stern, 1973; Ginsberg and Hillman, 1973).

3.3.3. Energetic Metabolism of Mammalian Ovum

The substrates utilized by the mammalian ovum for obtaining energy for metabolic processes have been determined biochemically, especially by successful culture of mouse and rabbit ova. Detailed data about the rat ovum are almost nonexistent, as culture of rat ova has so far been unsuccessful.

The following substances can serve as a source of energy for mammalian ova: glucose, fructose, galactose, lactate, pyruvate, oxaloacetate, phosphoenolypyruvate, and malate (Biggers and Stern, 1973). These sources are, of course, utilized to various extents. All mammalian cells utilize glucose as the basic energy substrate for metabolic reactions. As has been proved by Ishida (1968), Denker (1971), and Čech (1977a, b, c), glucose is acquired by the breakdown of glycogen which is present in the rat ovum in morphologically demonstrable particles whose amount is subjected to significant changes during cleavage. Details about demonstration and the metabolism of this important energy source are given in Chapter 3.5. It has been found (Whitten, 1957; Brinster, 1965a, b; Biggers et al., 1967) that one-cell and two-cell mouse ova do not divide in the presence of glucose alone as an energy source, while the eight-cell ovum develops further (Whitten, 1956; Brinster and Thomson, 1966). Glucose is thus unable to yield sufficient or better utilizable energy to the early developing ovum of the species under investigation and other sources must be used. By means of labeled glucose, it has been established that glucose is split into lactate and CO_2. Lactate is in turn utilized as an energy substrate (Brinster, 1967a, b; Wales, 1969). This finding might be connected to the changes in shape and arrangement of mitochondria during the eight-cell stage of cleavage, when significant changes take place in mitochondria which are doubtlessly reflected in metabolism (or, vice versa, changes in metabolism provoke morphologic changes). Similarly, fructose is utilized in the period between the two-cell and the eight-cell stages (Brinster and Thomson, 1966).

The other sources of energy, i. e., galactose, oxaloacetate, pyruvate, lactate, phosphoenolpyruvate, and malate, are substances which have been proved to enable mammalian ova to develop at a certain period and/or support this process, either independently or in combination. It is not certain whether some of them assert themselves in an unchanged form or after previous splitting into simpler compounds. It is possible that the mechanism of their utilization is not known (Whitten, 1957; Brinster, 1965b, 1970; Brinster and Thomson, 1966; Daniel, 1967; Biggers et al., 1967; Wales and Biggers, 1968; Wales and Brinster, 1968; Whittingham, 1969; Wales and Whittingham, 1970; Kramen and Biggers, 1971; Wales and Whittingham, 1973; Eppig, 1976). Some were proved to provoke abnormalities in pronuclei if added to the medium (Whittingham, 1969).

One of the indications of metabolism is oxygen consumption. Oxygen is especially necessary in the splitting of glucose and in the production of CO_2 which originates in the degradation of foodstuffs. From the comparison of oxygen consumption in the segmenting ova of the rat and of the mouse, it follows that oxygen consumption is increased in the two species. In mouse ova, the total consumption is only one-third as large as oxygen consumption in rat ova (Boell and Nicholas, 1948; Sugawara and Umezu, 1961; Gulyas and Daniel, 1967; Mills and Brinster, 1967); the maximum consumption is at the blastocyst stage. This observation could indicate increased anaerobic glucose utilization, which would also be supported by the established values of the activity of lactate dehydrogenase in the rat ovum (Brinster, 1965c; 1967b; Gibson and Masters, 1970; Engel and Kreutz, 1973; Poznakhirkina et al., 1975). This enzyme shows decreasing activity in the course of cleavage of the rat ovum; the minimum is reached at the blastocyst stage. The difference is also observed in the total amount of LDH in the rat and mouse ova. In the rat ovum, its total amount is one-third that of the mouse ovum, which supports the fact that aerobic glycolysis prevails in the rat ovum.

Carbon dioxide can originate in two ways: either as a product of aerobic glycolytic processes or as a product of the pentose shunt. Fridhandler et al. (1957), Fridhandler (1961), and Brinster (1968) found that glucose metabolism took place especially by means of the pentose shunt in rabbit ova up to the stage of the morula, after which there was a change to aerobic glycolysis. It is probable that the two processes take place in rabbit ova simultaneously, but in certain periods of cleavage either the one or other process prevails. In this case, an explanation for the change in glucose metabolism is offered by the onsetting morphologic changes in mitochondria during the eight-cell ovum and/or more advanced stages.

Energetic metabolism is closely connected to the content of ATP. The amount of ATP in mouse ova was determined by Quinn and Wales (1971) and Ginsberg and Hillman (1973). From those papers, it follows that the total abount of ATP decreases in the course of cleavage and, at the stage of the late blastocyst, is only one-third of the original value. It is necessary to bear in mind that these values denote the momentary state and/or the difference between synthesis and ATP decomposition, since the impetus for the synthesis is its consumption. That is why it is so difficult to correlate these findings with the changes in mitochondria which generally speak in favor of intense metabolism in this period of cleavage. The amount of ATP in rat ova has not been determined.

In the mouse and/or the rabbit ovum at different stages of cleavage, a number of ènzymes affecting the processes of energetic metabolism were proved biochemically (for references cf., Biggers and Stern, 1973). On the other hand, our information about enzymes included in the energetic events in the rat ovum is quite insufficient. So far only LDH activity has been studied (Brinster, 1967b; Gibson and Masters, 1970; Engel and Kreutz, 1973; Spielmann et al., 1974).

In reviewing the literature dealing with the metabolism of energetic substances, we have come across a basic drawback. Several substances and enzymes have been determined biochemically which participate in the process of acquiring energy. In no paper, however, do we come across the interpretation of the results from the point of view of topochemistry (with quite isolated exceptions), and cytochemical papers, either on the light or on the electron microscopic level, are missing altogether. There are a number of papers concerning the same problem, but employing different methods, so that their results are not comparable. In this situation, it is still very difficult to determine the changes in mitochondria due to metabolism and vice versa.

From enzymes participating in the reactions of energetic metabolism in the mammalian ovum, succinate dehydrogenase has been studied cytochemically on the light microscopic level (Ishida and Chang, 1965; Solter et al., 1972; Barańska et al., 1973; Parkening and Soderwall, 1973; Vivarelli et al., 1976); glycogen synthetase and phosphorylase in the segmenting rat ovum have also been studied (Čech, 1977a, d). On the ultrastructural level, succinate dehydrogenase has been proved by Šťastná (1977b, 1978b) in the mitochondria of the rat ovum. From another group of enzymes, organophosphate-resistant nonspecific esterase has been demonstrated by Trávník (1977b, 1978) in the mitochondria of the one-cell to eight-cell rat ovum. As for succinate dehydrogenase, the activity established is in contradiction to the expected result, as it has been found that in the more advanced stages its activity decreases. The results can, to a certain extent, be influenced by a change in the permeability of mitochondrial membranes of the more advanced stages (Šťastná, 1977c, personal communication).

From the data concerning the utilization of possible energy sources and the participation of enzymes, it is possible to conclude that the basic processes of energetic metabolism most probably take place in a similar way in the ova of the individual mammalian species. We cannot, however, attempt to generalize in this respect, as some observations (Fridhandler et al., 1957; Fridhandler, 1961; Brinster, 1968) of the metabolism of glucose in the rabbit and mouse ova show that the individual processes can have different courses in different animal species. This fact is also certified by the paper of Spielmann (1975), who found the utilization of energy sources in the mouse and rat ova in the period of the first division. In experiments with cultures of rat and mouse zygotes in various media, he found that rat zygotes underwent the first division in the medium with lactate or phosphoenolpyruvate, while mouse zygotes in the same medium did not divide. This effect is probably due to a different level of LDH in the two species (Brinster, 1965c; Epstein et al., 1971; Engel and Petzold, 1973). It is known that mouse zygotes are not capable of converting lactate to pyruvate, in spite of a high activity of LDH, which again is probably connected to the low concentration of NAD^+ in mouse zygotes (Sorensen, 1972; Zeilmaker et al., 1972; Cross and Brinster, 1973). The interpretation of the data obtained is difficult, since LDH activity in rat ova is much lower than in mouse ova of the same stage (Brinster, 1967b).

3.3.4. Conclusion

Energetic metabolism of the mammalian ovum has been only partly clarified. In biochemical papers dealing with the utilization of energetically rich substrates in ova of various mammalian species, it has been established that a tricarboxyl acid cycle takes place as well as the respiratory chain and oxidative phosphorylation, which result in energetically rich substances. These processes take place in a way similar to somatic cells. By the cytochemical proof of substrates as well as reaction products and enzymic systems, it has been demonstrated that the processes of energetic metabolism of the ovum also take place in mitochondria.

During rat ovum cleavage, the number, size, and inner structure of mitochondria change. The total volume of mitochondria does not essentially change in the course of cleavage. The change in their spheric or ovoid shape in the early stages to elongated, rod-like forms in the later stages of cleavage is an indication of the differentiation of this organelle. Correlations of cytochemical and morphologic studies show that changes in the fine structure, characterized by a different arrangement, course, and number of cristae, can be related to metabolic processes.

Current data on energy-obtaining processes and the manner in which energy sources are utilized by the mammalian ovum have so far been insufficient, and it is particularly necessary, therefore, to obtain ultracytochemical data on substances influencing energetic metabolism. Only then will it be possible to seriously correlate biochemical findings and express an opinion on metabolic processes with different courses.

3.4. Lysosomes and Their Importance for the Differentiation and Nutrition of the Segmenting Ovum

Jitka Šťastná

Lysosomes are cytoplasmic organelles present in both somatic cells and gametes of higher animal. Their main functional significance lies in their lytic capacity based on the presence of hydrolytic enzymes.

3.4.1. Intracellular Hydrolysis and Lysosomal Conception

Hydrolytic enzymes are of essential importance for the existence of all living systems. They play an important role in the intracellular turnover of macromolecules which is a continuous process in many kinds of eukaryonts. This turnover is based on enzymic degradation of macromolecules whose constituents are thereby made available for metabolic reuse. Hydrolytic enzymes also participate significantly in the turnover of cells and tissues of higher animals and in certain aspects of developmental remodeling. Some digestive enzymes are released into the outer environment and then take part in nutritive processes, the turnover of extracellular materials of connective tissues, or other special processes. Thus, degradative enzymes participate in fertilization, in the antimicrobial defense system, and in a number of other processes taking place both under normal and pathologic conditions. A list of hydrolases identified so far and their substrate specifity is given by Holtzman (1976).

On the cell level, hydrolytic enzymes are contained in the "intracellular digestive system" (de Duve and Wattiaux, 1966). In this system, the central role belongs to lysosomes, membrane-limited structures containing chracteristic hydrolytic enzymes, most of which have acid pH optimum. Lysosomes are characterized by considerable morphologic and functional variety. They are classified into two basic groups; primary and secondary lysosomes.

Primary lysosomes are structures containing hydrolases originating from biosynthesis which have not yet participated in the digestive processes. In most of the cell types, they are morphologically difficult to identify. Cytochemical papers state that they are identical with some small vesicles formed by both the Golgi complex or Golgi complex-associated smooth endoplasmic reticulum (GERL) (Holtzman, 1976). The main function of primary lysosomes is the transport of hydrolytic enzymes into secondary lysosomes.

Secondary lysosomes are a heterogeneous group from the morphologic point of view. Their appearance varies greatly, depending on the mechanisms of origin, properties of the digested material, the phase of the digestive processes, the content of indigestible remnants, etc. According to the origin of material digested in secondary lysosomes, they are divided into the heterophagic and the autophagic lines.

In heterolysosomes, material incorporated by endocytosis is digested. The structures arising from endocytosis — phagosomes (Straus, 1967) — do not initially contain hydrolytic enzymes at first but obtain them secondarily by fusion with lysosomes. The fusions between lysosomes and endocytotic vacuoles are one of their basic properties (Gordon et al., 1965; de Duve and Wattiaux, 1966).

Autophagic vacuoles are a morphologic manifestation of autophagy, characterized by the presence of the limiting membrane (or sometimes two membranes), which contain — besides ground cytoplasm — various cell organelles and structures at different degrees of degradation. The occurrence of autophagic vacuoles, their genesis, and functional importance were studied in detail by de Duve and Wattiaux (1966), Ericsson (1969), Holtzman (1976).

Multivesicular bodies which are known to be engaged both in heterophagic and autophagic processes constitute a special group of lysosomes (Holtzman, 1976). They occur in a wide range of cell types. From the morphologic aspect, these bodies have a variable appearance, a limiting membrane and contents of small elementary vesicles are characteristic. Multivesicular bodies of various cell types can differ in other details. They are ranked among lysosomes because some of them exhibit acid phosphatase (de Duve and Wattiaux, 1966) and aryl sulphatase (Holtzman, 1969, 1971).

All of secondary lysosomes forms mentioned change into residual bodies in which their origin can no longer be distinguished (de Duve and Wattiaux, 1966). The above change is due either to continuing digestive processes or frequent mutual fusions. These bodies often contain parallely arranged membranes or their fragments, lipid droplets, and electron dense granules in the amorphous matrix. Residual bodies can still be enzymically active or can already lack hydrolytic enzymes (Daems et al., 1969).

3.4.2. Forms and Representation of Lysosomes in the Segmenting Ovum

Structures of lysosomal nature have been determined in the rat ovum on the basis of morphologic observations and cytochemical proof of acid phosphatase (Šťastná, 1974a, b, c, 1977a) and nonspecific esterase (Trávník, 1977b, 1978). It follows that both primary and secondary lysosomes occur in the preimplantation period.

3.4.2.1. Primary Lysosomes

Primary lysosomes in the rat ovum have the shape of small smooth vesicles with a size of 0.05 - 0.1 μm and exhibit acid phosphatase activity. Such vesicles are particularly numerous in the Golgi region but also occur elsewhere in the cytoplasm. As is the case in somatic cells, these primary lysosomes in the rat ovum are probably formed by the Golgi complex. This is witnessed by the proof of acid phosphatase in the Golgi complex of the two-cell stage and the blastocyst (Schlafke and Enders, 1973). In the same mammalian species, Šťastná (1977a, 1978a) found the presence of acid phosphatase activity both in the cisternae and vesicles of the Golgi complex during the whole preimplanation period. In addition, in the structures of the Golgi complex of the two-cell stage of the rat, further lysosomal.hydrolase, the nonspecific esterase has been proved (Kukletová et al., 1974; Trávník, 1977b, 1978).

Primary lysosomes split from the Golgi complex are probably engaged in the transport of hydrolytic enzymes into secondary lysosomes in a way similar to that supposed for a number of different somatic cells. Like in somatic cells in the rat ovum, secondary lysosomes in the neighborhood of the Golgi complex are particularly numerous, or small vesicles with acid phosphatase activity, morphologically identical to the Golgi vesicles, are often located close to secondary lysosomes.

The second type of structures doubtlessly participating in the transport of hydrolytic enzymes in the rat ovum are tubules and vesicles of smooth endoplasmic reticulum. In almost all of the structures of smooth endoplasmic reticulum, acid phosphatase activity (Šťastná, 1977a) has been proved in the early stages of cleavage, while organophosphate-sensitive esterase activity has been proved in most of the structures and organophosphate-resistant esterase activity (Trávník, 1977b, 1978) in only part of the structures.

Characteristic complexes of tubules and vesicles of smooth endoplasmic reticulum of the early stages are topographically closely related to the Golgi complex. Secondary lysosomes are especially numerous in their neighborhood. Mutual relations of the above structures can be interpreted to mean that the vesicles of smooth endoplasmic reticulum transport hydrolytic enzymes into the Golgi complex. This transport is supported by the cumulation of the reaction product in the proof of acid phosphatase in the structures of the Golgi complex in a substantially grater amount than in the vesicles and tubules of smooth endoplasmic reticulum. It is also possible that, like in some types of somatic cells (Holtzman, 1976), hydrolytic enzymes are transported directly into the lysosomes by the vesicles separated from smooth endoplasmic reticulum. Besides close topographic affinity of these vesicles to lysosomes, especially to multivesicular bodies, this idea is also supported by the fact that the decrease in the quantity of smooth endoplasmic reticulum with acid phosphatase activity is accompanied by the increase in the number of enzymically active secondary lysosomes.

A large number of tubules and vesicles of smooth endoplasmic reticulum represents a storage of hydrolytic enzymes in the early stages of cleavage, since new hydrolytic enzymes cannot be formed due to the absence of the proteosynthetic apparatus (cf., 3.2.). During cleavage, the amount of enzymically active structures of smooth endoplasmic reticulum decreases and occur only rarely in the morula and blastocyst. In these stages, acid phosphatase is present – in addition to lysosomes and the structures of the Golgi complex – in granular endoplasmic reticulum (Šťastná, 1977a). Its presence in the cisternae of granular endoplasmic reticulum can be considered an indication of the renewed synthesis of hydrolytic enzymes.

Some evidence supports the fact that cortical granules could also belong to primary lysosomes in mammalian ova. These granules were observed in hamster ova (Austin, 1956) for the first time in the light microscope by means of phase contrast and were studied in the ova of many mammalian species under the electron microscope (Austin, 1961; Szollosi, 1962, 1967; Hadek, 1963; Adams and Hertig, 1964; Hope, 1965; Weakley, 1966; Zamboni and Mastroianni, 1966a; Baca and Zamboni, 1967; Wischnitzer, 1970; Dvořák, 1971; Norberg, 1972; Šťastná, 1974b, c; Kang and Anderson, 1975).

Cortical granules of mammalian ova are formed by the Golgi complex (Zamboni and Mastroianni, 1966a; Baca and Zamboni, 1967; Szollosi, 1967; Zamboni, 1970; Kang, 1974; Selman and Anderson, 1975). They are uniform in appearance (round or oval), and their size is about 0.3 μm. They are limited by a membrane and contain homogeneous, intensely electron dense material. During fertilization, they disappear from ovum cytoplasm in the course of the cortical reaction. This reaction begins at the place of sperm penetration and consists in releasing the content of cortical granules into the perivitelline space. In the rat ovum, several granules sometimes fuse laterally by forming spacious cortical caverns under the surface of the ovum (Szollosi, 1967).

The chemical composition of cortical granules of mammalian ova is not yet exactly known. The presence of proteins, glycoproteins, and acid mucopolysaccharides has been demonstrated (Szollosi, 1967; Guraya, 1969; Fléchon, 1970; Selman, 1974; Selman and Anderson, 1975). Acid phosphatase, when present in cytoplasmic bodies generally considered to be a marker of lysosomal nature in mature cortical granules, has not yet been demonstrated (Anderson, 1972; Kang, 1974; Šťastná, 1974b, c), with the exception of acid phosphatase in the cortical granules of rabbit ova (Hadek, 1969). Kukletová et al. (1974) did not find another hydrolytic enzyme — nonspecific esterase — in cortical granules of rat oocytes. Acid phosphatase activity has, however, been found in some immature cortical granules of rat oocytes by Kang (1974) and in the contents of cortical granules of freshly fertilized rat ova immediately after their breakdown by the cortical reaction, when enzymic activity was demonstrated in some cortical caverns and occasionally also in the material of single releasing granules (Šťastná, 1974b, c). These positive results of the reaction to acid phosphatase in the material of cortical granules of rat ova support their lysosomal character. Owing to the origin of these granules from the Golgi complex, whereby their uniform morphology does not support the fact that they have digested some material, it is possible to assume that cortical granules are a special form of primary lysosomes whose hydrolytic enzymes act outside the cell.

It is now generally accepted that the block to polyspermy in fertilized mammalian ova results from the breakdown of cortical granules (Gwatkin, 1976). The contents of cortical granules act upon the zona pellucida which changes its properties (zona reaction) and becomes impermeable to further sperms (Braden et al., 1954; Austin, 1961; Stegner, 1967; Gwatkin et al., 1973). In cortical granules of hamster ova, Gwatkin et al. (1973) observed a trypsin-like protease which prevents the binding of spermatozoa to the zona pellucida. Szollosi (1967) and Hadek (1969) state that the material of cortical granules — or possibly hydrolytic enzymes contained in it — can disturb excess sperms which have penetrated into the perivitelline space, thus preventing them from penetrating into the ovum.

3.4.2.2. Secondary Lysosomes

Secondary lysosomes are present in all stages of the segmenting rat ovum and in the blastocysts. Their amount increases during cleavage. In the population of secondary lysosomes, two morphologically distinct groups of cytoplasmic bodies occur: multi-vesicular bodies and autophagic vacuoles. Both types converge into such forms of secondary lysosomes that it is impossible to distinguish their origin and are ultimately converted into residual bodies. Secondary lysosomes are doubtlessly present in ova of all mammalian species. Their denotation, however, varies greatly, with the exception of multivesicular bodies. Thus, the bodies of that type in rat ova have also been denoted as inclusion bodies (Mazanec, 1965; Schlafke and Enders, 1967) or degradative bodies (Enders, 1971), in mouse ova as multilaminar bodies (Wischnitzer, 1970), secondary lysosomes, or telolysosomes (Gianguzza and Mulnard, 1972), in rabbit ova as dense lysosome-like bodies (Merchant, 1970), and in mouse, guinea pig, and rabbit oocytes as dense bodies (Anderson, 1972). This category probably also includes lamellar bodies in the oocytes of the hedgehog (Sydow, 1968) or compound aggregates in human oocytes (Hertig and Adams, 1967).

Multivesicular Bodies

Multivesicular bodies are structures generally occurring in mammalian ova. They have been observed in mouse ova (Yamada et al., 1957; Calarco and Brown, 1969; Maraldi and Monesi, 1970; Anderson, 1972), in rat ova (Sotelo and Porter, 1959; Odor, 1960; Franchi and Mandl, 1962; Izquierdo and Vial, 1962; Mazanec and Dvořák, 1963; Mazanec, 1965; Schlafke and Enders, 1967; Stegner, 1967; Dvořák, 1971), in guinea pig ova (Anderson and Beams, 1960; Adams and Hertig, 1964; Anderson, 1972), in rabbit ova (Blanchette, 1961; Krauskopf, 1968; Merchant, 1970), in hamster ova (Odor, 1965), in monkey ova (Hope, 1965), and in human ova (Wartenberg and Stegner, 1960).

Multivesicular bodies are extraordinarily numerous in rat ova. They occur most frequently in the unfertilized and fertilized ova, where they comprise 2.1% of the cytoplasmic volume; their number decreases during cleavage. In the eight-cell stage, they account for only 0.4% of the cytoplasmic volume and in later stages occur only rarely (Dvořák et al., 1977). The decrease in number of multivesicular bodies during cleavage was also established in mouse ova (Calarco and Brown, 1969) and rabbit ova (Merchant, 1970). Multivesicular bodies in the rat ovum are, like all forms of secondary lysosomes, very polymorphous structures, their appearance depending particularly on the amount of elementary vesicles and amorphous material of variable density they contain. According to morphologic properties, they can be divided into two basic types; light bodies with low contents of elementary vesicles with a size of 10 - 50 nm in electron microscopically empty environment and dark multivesicular bodies with a dense matrix and containing elementary vesicles and vacuoles with a size of 20 - 140 nm. Bodies with a nucleoid, an area of dense matrix located centrally probably constitute a transitional form. In addition to granules of dense material difficult to identify, some multivesicular bodies also contain membrane fragments. The presence of these structures is considered to be a sign of digestion which is in process or has taken place, and such bodies probably represent transitional forms of multivesicular and residual bodies. The transformation of multivesicular bodies into residual bodies is evidenced by the decrease in the number of multivesicular bodies accompanied by the increase in the number of residual bodies.

Acid phosphatase has been discovered in only some multivesicular bodies, i.e., in one ovum it appeared in positively reacting bodies next to negative ones. Similar results.

were obtained by Stegner (1967) in multivesicular bodies of rabbit ova. Light bodies did not contain reaction products at all or in only small quantities. Positive reaction was, however, found regularly in dark bodies, in bodies with the nucleoid, and in transitional forms of multivesicular and residual bodies. Nonspecific esterase has been established in some multivesicular bodies of rat oocytes (Kukletová et al., 1974), but using another method, Trávník (1977b, 1978) did not find its presence in these structures. In some multivesicular bodies of the early stages of the segmenting rat ovum, alkaline phosphatase activity has been detected (Šťastná, 1979).

Based on the occurrence of multivesicular bodies particularly in the cortical cytoplasm of the ovum and their close relationship to pinocytotic vesicles, it can be stated that they can be derived from endocytotic vesicles and vacuoles and thus can have the character of heterolysosomes (Šťastná, 1974a). This type of character of multivesicular bodies of mammalian ova is manifested by the presence of injected ferritin in multivesicular bodies in rabbit ova (Hadek, 1969), horseradish peroxidase and acid phosphatase in multivesicular bodies in the ova of the guinea pig, mouse, and rabbit (Anderson, 1972), or exogenous peroxidase in multivesicular bodies of rabbit blastocysts (Hastings II and Enders, 1974a). On the other hand, there is no evidence supporting the possibility of their autophagic origin.

Due to the lysosomal character of multivesicular bodies and their participation in the uptake of exogenous macromolecules, it is possible to suppose that these structures have a certain importance for the nutrition of the ovum. Doubtlessly, they already assert themselves in the ovarian oocytes, as they are represented in high numbers and exogenous proteins and hydrolytic enzymes have also been demonstrated (Anderson, 1972). Based on the occurrence of multivesicular bodies especially in the early stages of cleavage, it seems probable that with the progress of cleavage the uptake and degradation of macromolecules from outside by means of multivesicular bodies loses its importance.

Autophagic Vacuoles

Data concerning the occurrence and function of autophagic vacuoles in mammalian ova have so far been rather limited. The presence of autophagic vacuoles in rat blastocysts is mentioned by Schlafke and Enders (1973), who have also demonstrated acid phosphatase.

Autophagic vacuoles occur in the course of the whole preimplantation development of the rat ovum; their representation and morphological properties, however, change significantly with the proceeding development. In the early stages of cleavage, up to the four-cell ovum, autophagic vacuoles of small dimensions occur (0.1 - 0.5 μm), in which small areas of ground cytoplasm with accumulated glycogen particles are separated. From the four-cell stage onward, in addition to small autophagic vacuoles, vacuoles 1 - 2 μm start appearing, which contain mitochondria, structures of endoplasmic reticulum, small vesicles, ribosomes, lamellar structures, and glycogen, in the cytoplasm of the blastomeres. The maximum occurrence of autophagic vacuoles containing still intact or slightly damaged cell components has been found in the eight-cell stage and in the morula; in blastocysts, newly formed autophagic vacuoles have been observed less frequently.

From the study of initial stages of autophagic vacuoles, it follows that vesicles and tubules of smooth endoplasmic reticulum, cisternae of granular endoplasmic reticulum, Golgi structures, but also secondary lysosomes – particularly multivesicular and residual bodies – participate in the separation of cytoplasmic regions (Šťastná, 1977a). All these forms of the genesis of autophagic vacuoles have been observed even in different kinds of somatic cells (Ericsson, 1969; Dvořák, 1974b).

In most of the autophagic vacuoles of the two types mentioned, acid phosphatase activity has been proved. Vacuoles with positive and negative reactions have been jointly present in all cells of the individual stages. Smooth vesicles with acid phosphatase activity, which could be primary lysosomes, have been regularly observed close to autophagic vacuoles. In most of the autophagic vacuoles, organophosphate-resistant esterase activity has also been established (Trávník, 1977b, 1978). In some vacuoles, even alkaline phosphatase activity has been proved (Šťastná, 1979).

The presence of numerous autophagic vacuoles during rat ovum cleavage is not surprising, as it is known that these structures occur in cells under normal circumstances. They accompany differentiation processes in high numbers. Small autophagic vacuoles with accumulated glycogen of the early stages probably, particularly serve the degradation of glycogen (cf., 3.5). The degradation of cell organelles and other cell components by autophagy takes place to a great extent in the four-cell but especially in the eight-cell stage and in the morula, i. e., the stages characterized by a gradual renewal of proteosynthetic activity (cf., 3.2.). It is probable that in this period the need to degrade worn cell organelles asserts itself. This hypothesis is confirmed above all by frequent observations of mitochondria typical of early stages of cleavage inside autophagic vacuoles. This phenomenon is evidently connected to the morphologic reconstruction of mitochondria which occurs at these stages (cf., 3.3). On the other hand, it is possible that the sense of autophagy is to obtain energy or simpler chemical substances for new synthetic processes.

Residual Bodies

In the segmenting rat ovum, residual bodies constitute an important group of lysosomes. They are represented in relatively small numbers in the one-cell and two-cell stages; their number increases in the four-cell ovum and are then very numerous up to the blastocyst stage. A similar trend in the incidence of these structures during the cleavage of the mouse ovum is given by Gianguzza and Mulnard (1972). In the early stages of cleavage, residual bodies of the rat ovum have about the same size as multivesicular bodies (0.3 - 0.5 μm). They are limited by a membrane and generally contain a large amount of dense amorphous material, membrane fragments, and vacuoles of different sizes. Residual bodies of these early stages probably originate from multivesicular bodies. They show regularly intense positive reactions to acid phosphatase (Šťastná, 1974a; 1977a), organophosphate-resistant esterase (Trávník, 1977b; 1978), and alkaline phosphatase (Šťastná, 1979).

In later stages of cleavage and in the blastocysts, relatively large residual bodies (average size 1 - 2 μm) accumulate in the cytoplasm of blastomeres and in embryoblast and trophoblast cells. They are limited by a membrane and mostly contain large amounts of membrane fragments, often lamellarly arranged, but also granules of different sizes, vesicles, vacuoles, materials of different densities difficult to define, and often lipid droplets. Great differences have been observed in the intensity of the reaction to acid phosphatase in these bodies. Bodies with intense reactions and those in which the reaction product was either missing altogether or only minimal were observed. In most residual bodies, organophosphate-resistant esterase activity (Trávník, 1977b, 1978) was found. A weak alkaline phosphatase activity (Šťastná, 1979) was only established in individual cases.

The above changes in the morphology of residual bodies are probably due to the fact that in later phases of cleavage these bodies develop from autophagic vacuoles or possibly by the fusion of various forms of lysosomes. Residual bodies represent the last stage of development of secondary lysosomes of the rat ovum. From the results of cytochemical reactions to hydrolytic enzymes, it follows that most of these bodies are still active secondary lysosomes, while part of them are already inert residual bodies.

3.4.3. Conclusion

In the course of the preimplantation development of the rat ovum, lysosomes are the most frequently represented cytoplasmic structures. They assert themselves significantly both in nutritive and diferentiation processes, whereby cortical granules play a special role. Primary lysosomes, in the form of small smooth vesicles exhibiting acid phosphatase and nonspecific esterase activity, are particularly numerous in the early stages of cleavage. They are derived from the Golgi complex and probably also from the smooth endoplasmic reticulum. A reason for ranking cortical granules of the rat ovum with primary lysosomes is the proof of acid phosphatase activity in some immature cortical granules and in the material of granules after their lysis during the cortical reaction. Hydrolytic enzymes of those granules assert themselves extracellularly.

Secondary lysosomes are very numerous throughout the preimplantation development. In the early stages, they are chiefly represented by multivesicular bodies in which material transported into the ovum by endocytosis is probably degraded. In the course of the whole cleavage, autophagic vacuoles occur, starting from the four-cell and particularly the eight-cell stage, and cell organelles are degraded to a great extent by autophagy. Autophagic vacuoles accompany differentiation processes in general; their main task is to degrade worn cell components or possibly to obtain simpler substances or energy for new synthetic processes. With the advance of digestive processes, both multivesicular bodies and autophagic vacuoles initially converge into secondary lysosomes in which it is no longer possible to differentiate their origin and then ultimately into residual bodies. The incidence of these forms gradually increases during cleavage up to the blastocyst stage.

3.5. Occurence of Stored Material and Its Metabolism in Segmenting Ova

Svatopluk Čech

Stored material in the ova of vertebrates is summarily termed egg yolk (vitellus, lekithos). From the point of view of its quantity mammalian ova are ranked among oligolecithal ova. Chemically, yolk is material that is pronouncedly heterogeneous. Besides proteins, it also contains lipids and saccharides, especially glycogen. The occurrence and changes of each of the chemically defined substance groups during cleavage will be dealt with in a separate section.

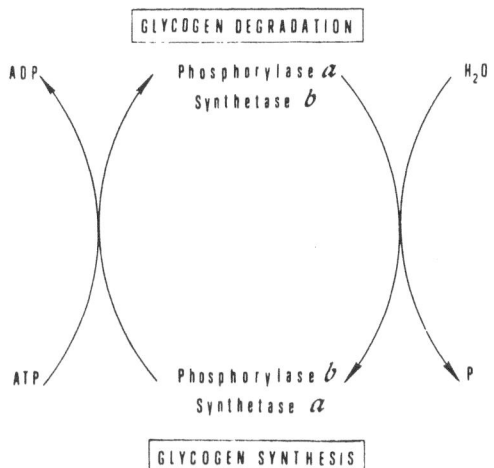

GLYCOGEN DEGRADATION

ADP   Phosphorylase a H₂O
       Synthetase b

ATP   Phosphorylase b P
       Synthetase a

GLYCOGEN SYNTHESIS

Fig. 62. Scheme of the interaction between active and inactive phosphorylases and synthetases in the processes of glycogen degradation and synthesis.

3.5.1. Glycogen

Glycogen content in animal cells is a momentary expression of the level of two antagonistic processes, i.e., glycogenesis and glycogenolysis. In the former, the key role is played by glycogen synthetase, in the latter α-glucan phosphorylase. Both enzymes exist in two interconvertible forms: active (phosphorylase a and glycogen synthetase a) and inactive (phosphorylase b and glycogen synthetase b). The conversion of one form into the other and vice versa is controlled by means of specific kinases (phosphorylation) and specific phosphatases (dephosphorylation). The interaction between the two enzymatic systems is shown in Figure 62 according to Hers et al. (1970). Of basic importance is the fact that the occurrence of the active form of one of the enzymes assumes the existence of the inactive form of the other and vice versa. For these reasons, it is necessary to take a complex approach toward the study of glycogen, always in close connection with the critical examination of the activity of key enzymes of glycogen metabolism.

3.5.1.1. Biochemical and Cytochemical Data on the Occurrence and Changes of Glycogen and Some Related Enzymes During the Cleavage of Some Mammalian Ova

For the sake of clarity and due to the fact that possible species differences cannot be excluded, accessible data are given separately for every species. In the mouse, the results of quantitative biochemical analysis of one-cell unfertilized ova have shown that carbohydrates constitute about 16% of the dry weight (Loewenstein and Cohen, 1964). Glycogen and the changes in its content during the cleavage of mouse ova were studied by Stern and Biggers (1968). They found that the amount of glycogen grew continuously up to the eight-cell stage when it reached its maximum (from an initial value of 0.14 ng/one-cell unfertilized ovum to 2.24 ng/eight-cell ovum). In the blastocyst stage, the decrease in glycogen was evident. Further quantitative biochemical data were published by Ozias and Weitlauf (1971), Snyder et al. (1971), and Ozias and Stern (1973). The values established by these authors agree in principle with the results achieved by Stern and Biggers (1968). Cytochemical detection of glycogen by means

105

of the PAS method in segmenting mouse ova was carried out by Thomson and Brinster (1966). The findings of these authors on the whole correspond well to biochemical results. The authors state that very intense PAS-positive reaction was found in two-cell ova to the morula, while in the blastocysts the content of PAS-positive material was significantly lower, the same as in one-cell ova. Of the enzymes connected with glycogen metabolism, only glycogen synthetase was studied in mouse ova; its biochemical determination was carried out by Stern (1970). He found that its activity during cleavage increased and culminated in the eight-cell stage. Further development of the ovum was accompanied by a rapid decrease in enzymatic activity, so that blastocysts contained only 5% of the activity established in the eight-cell ova.

In the segmenting hamster ova, the presence of glycogen was confirmed by Mc-Reynolds and Hadek (1972b) who employed the PAS method and diastase digestion. They state that there were no significant differences between the individual stages of cleavage in the occurrence of PAS-positive, diastase labile material. According to the findings of Parkening and Soderwall (1974), only one-cell ova of this species show intense PAS-positive reaction; starting with the two-cell ovum, the intensity gradually decreases up to the late blastocyst stage. Cytochemical detection of α-glucan phosphorylase and glycogen synthetase in segmenting hamster ova was carried out by Ishida (1968). He found that up to the eight-cell stage the activity of the former enzyme was constant but disappeared completely by the blastocyst stage. The activity of glycogen synthetase was highest at the beginning of cleavage and then gradually decreased; in the blastocysts, the result of the examination was negative.

The occurrence and changes of glycogen content with the aid of cytochemical examination in the segmenting pig and rabbit ova were dealt with by Ishida (1963). He found that, while in the pig ova glycogen content had not changed during cleavage, its amount in rabbit ova had increased. The presence of glycogen in all preimplantation stages of the rabbit was also determined by Denker (1970a, b). More significant changes in glycogen content − in the sense of substantial decreases − were observed as late as the blastocysts stage, which was also independently verified by Lutwak-Mann (1971). As for the enzymes connected with glycogen metabolism in segmenting rabbit ova, the cytochemical papers by Ishida (1968) and Denker (1971) offer a good starting point. According to the findings made by Denker (1971), α-glucan phosphorylase is present at all stages of cleavage, including the blastocyst stage, whereas Ishida (1968) no longer found it in the blastocysts. Glycogen synthetase was proved only in the earlier stages and was always missing in the blastocysts (Ishida, 1968).

The occurrence of glycogen in the course of cleavage of rat ova was studied by Ishida (1954), Mulnard and Dalcq (1955), Baeckeland (1975a, b), and Čech (1977a). Mulnard and Dalcq (1955) and Baeckeland (1975a, b) maintain that glycogen is not subjected to major changes during the cleavage of the rat ovum, while Ishida (1954) and Čech (1977a) established its decrease. Cytochemical detection of the enzymes of glycogen metabolism in rat ova was carried out by Čech (1977a, d). He found that α-glucan phosphorylase was present especially in the active form (as phosphorylase a). Up to the eight-cell stage, it showed constant and high activity; a significant decrease was observed in the blastocysts, some of which still showed a weak positive reaction while others did not. In the case of glycogen synthetase, he showed that segmenting rat ova contained only the glucose-6-phosphate-dependent inactive form (synthetase b). The maximum of enzymatic activity was shown in the one- and two-cell ova; in later stages, the reaction was weak and in the blastocyst stage always negative.

The analysis of the above facts, in spite of possible partial discrepancies, shows that there is evidently some relationship between the changes in glycogen content and the activity of the respective enzymes. This means that in the segmenting ova, the same as in somatic cells, its actual amount depends on the intensity of synthetic and degradation processes. The prevalence of the former or of the latter, reflected in the differences in the occurrence of the forms and activities of glycogen synthetase and α-glucan phosphorylase in cytochemical examinations, decides whether the level of glycogen will increase or decrease during the cleavage of the ovum. Seen from this point of view, the results of cytochemical investigations in rabbit, hamster, and rat ova support the fact that the amount of glycogen decreases during cleavage, that it is metabolically mobilized, and subsequently degraded. In all three species, the degradation enzyme, α-glucan phosphorylase, was present in the active form almost throughout the whole cleavage, whereas glycogen synthetase was found — only in the inactive form — merely at its beginning (Ishida, 1968; Čech, 1977a). As for glycogen itself and its changes during the cleavage of ova of the species mentioned, according to some authors its level remains the same, while according to others it decreases. The reason for these differences is the fact that mostly ova in toto were used to obtain the results of the PAS method, which does not guarantee as much accuracy as the evaluation of the reaction in sections (Čech, 1977a). To be able to draw the same conclusions in mouse ova, information is necessary on α-glucan phosphorylase, which has not yet been studied. It appears, however, that the situation in this species will be different. In mouse ova, the level of glycogen rises during cleavage and reaches its maximum in the eight-cell stage (Stern and Biggers, 1968). A similar trend in changes has also been established in the case of glycogen synthetase (Stern, 1970).

3.5.1.2. Submicrocscopic Occurrence of Glycogen in Rat Ova

Submicroscopic occurrence of glycogen in mammalian ova has been studied only occasionally. Exact results, based on the ultracytochemical method introduced by Thiéry (1967) for the detection of glycogen, have so far only been obtained in rat ova. The first preliminary examinations were carried out by Dvořák and Trávník (1972) and Dvořák et al. (1973) and continued by Čech (1977a, b, c). Detailed descriptions of glycogen distribution in the individual stages are given in section 2.

Ultracytochemical examinations have not only proved the validity of light microscopic findings about the decrease in glycogen amount during the cleavage of rat ovum but have also revealed some new facts, i.e., segregation and accumulation of glycogen in some cell organelles and extracellular deposits of polysaccharide found in eight-cell stages and early blastocysts.

Cytoplasmic glycogen segregation starts before the first mitosis. The process culminates in the two-cell ovum and decreases starting with the next stage of cleavage. It is manifested by glycogen particles gathering into aggregations of various sizes, usually located close to larger sacs of smooth endoplasmic reticulum. In conjunction with the participation of these sacs, glycogen aggregations separate from the ground cytoplasm until they are finally incorporated. The average size of these smooth sacs with glycogen is 0.4 - 0.6 μm. Owing to the autophagic mechanism of origin, these structures can be considered a form of small autophagic vacuoles. Multivesicular bodies also participated, although rarely, in the separation, mostly partial, of some aggregations of glycogen particles. Starting with the eight-cell ova, smooth sacs with glycogen can no lon-

ger be observed. A characteristic feature of this stage, as well as of the early blastocysts, is the accumulation of glycogen particles in large autophagic vacuoles with remnants of cell organelles. Mere traces of polysaccharides were contained in autophagic
vacuoles of late blastocysts. Solitary glycogen particles occurred in small numbers of
residual bodies throughout the cleavage.

In looking for an answer to the question of what the above findings mean it is necessary to start from the fact that both smooth sacs with glycogen and autophagic
vacuoles with remnants of cell organelles and glycogen as well as multivesicular bodies
are organelles of lysosomal nature. On the basis of submicroscopic location of acid
phosphatase, Šťastná (1974a, b) confirmed this fact in the segmenting rat ovum. The
digestion of glycogen in lysosomes of different somatic cells was proved biochemically
(Hers, 1963; Badhuin et al., 1964) and also on the basis of electron microscopic observations (Jézéquel et al., 1965; Phillips et al., 1967; Kotoulas et al., 1971; Kotoulas and
Phillips, 1971; and others). Rosenfeld (1964) is of the opinion that the hydrolytic
degradation of glycogen in lysosomes leads to a quick release of free glucose and that
their physiologic participation in the breakdown of glycogen is a reflection of this requirement.

Concerning the problem of the participation in phosphorolytic breakdown by glycogen located in the ground cytoplasm, catalyzed by means of α-glucan phosphorylase,
and the breakdown of glycogen inside the lysosomes, it is possible to state in general
that, according to present opinion (reviewed by Huijing, 1975), the latter way is considered the less important. The main reason is that the volume of lysosomal glycogen
in comparison to that of glycogen located in the ground cytoplasm is relatively small.
As has been proved by the results of morphometric studies carried out by Dvořák et
al. (1974, 1977), not even the segmenting rat ova are an exception in this respect. The
volume of secondary lysosomes, mostly autophagic vacuoles, determined by the
authors, was only 0.42% in the eight-cell ova, when it reached its maximum. It is thus
possible to conclude that lysosomal degradation of glycogen in the segmenting rat
ovum represents a secondary, probably necessary complementary way, in spite of the
fact that it dominates the ultrastructural picture. The role of free glucose in the segmenting ovum is not known.

Among the more interesting findings, extracellular aggregations of glycogen surrounded by a membrane have been observed. In the four- and eight-cell stages, they
have been located between the blastomere surface and the zona pellucida. The number
of extracellular aggregations of glycogen particularly increases in the early blastocyst.
They are found in intercellular spaces and after the formation of common blastocyst
cavity chiefly near the cell surface limiting it. In some cases, similar aggregations of
glycogen granules were observed in the cell cytoplasm, in the immediate vicinity of the
cell membrane. Even though an explanation has been offered that their extrusion takes
place, there is no direct evidence so far. The importance of extracellular glycogen aggregations is not known. It is, however, not possible to exclude the fact that they
might serve as a source of free glucose for the blastocyst fluid.

3.5.1.3. Some Notes on the Mechanisms of Glycogen Metabolism Control in Mammalian Ova

Only vague theories exist in the mechanisms controlling glycogen metabolism in the
preimplantation stages of mammals. Based on the fact that α-glucan phosphorylase,

the same as glycogen synthetase, has been demonstrated in some mammalian ova, it can be stated that they will be realized by means of interconversions between their active and inactive forms. In the segmenting ova, it is necessary also take into consideration the influence of the tubal milieu, which changes dynamically depending on the hormonal state of the female (Stone and Hammer, 1975). Qualitative changes in the luminal fluid of the tuba uterina seem to be determining factors under some circumstances. A controlling or at least modifying influence on the total amount of glycogen in tubal ova can be effected by the lack, or, on the other hand, excess of precursors in the tubal environment, according to culture experiments carried out by Ozias and Stern (1973). Of interest are the findings made by Snyder et al. (1971), who described a gradual increase in glycogen content in the ova of intact or hypophysectomized mice after ligation of one oviduct. The increase is six times higher than in control ova (on the nonligated side), while the content of glucose in the tubal fluid of the ligated and nonligated oviduct remain the same.

Besides the tubal milieu, as shown by further experimental data, glycogen content in the segmenting ova can also be influenced by ovarial hormones. In mouse blastocysts, this was demonstrated by Ozias and Weitlauf (1971) and Snyder et al. (1971). These authors found that ovariectomy, the same as progesterone administration to fertilized females, provoked a decrease in glycogen level in ova, whereas estrogen administration resulted in its increase. The progesterone effect may possibly be explained by its effect on adenyl cyclase whose level it increases (Rosenfeld and O'Malley, 1970).

Glycogen synthesis from glucose was repeatedly confirmed only in mice ova (Brinster, 1969; Ozias and Stern, 1973; Pike and Wales, 1975). It seems, however, that even simpler substances can assert themselves as further precursors. In mouse ova they are, e.g., pyruvate or lactate (Biggers, et al., 1967; Brinster, 1969; Ozias and Stern, 1973; Pike and Wales, 1975; and others).

On the basis of the results of cytochemical determinations of α-glucan phosphorylase and glycogen synthetase activities, it is possible to presume that the degradation of glycogen occurs in the segmenting ova of the hamster, rabbit, and rat (Ishida, 1968; Denker, 1971; Čech, 1977a). In rat ova (Čech, 1977a) it has been established that its degradation occurs in two ways: phosphorolytically and by hydrolytic breakdown in lysosomes. In the former case, glucose-1-phosphate results, in the latter free glucose. As far as the importance of glucose-1-phosphate is concerned, it is possible to theoretically consider its utilization inside the Embden-Meyerhof metabolic pathway under anaerobic or aerobic conditions and/or its direct oxidation inside the pentose shunt. Lysosomal glycogen degradation may be a secondary process whose determination is not known in detail.

3.5.2. Lipids

From the chemical point of view lipids are rather diverse chemical substances whose only common feature is the prevalence of hydrophobic groups inside the molecule. They are usually divided into neutral fats, waxes, phospholipids, carotenoids, and steroids. They occur in cells not only as integral components of cell structures but also in the form of stored material.

The occurrence of lipids in mammalian ova, particularly in oocytes, has been dealt with by a number of papers, both histochemically and electron microscopically (see review by Trávník, 1977a). Even if it might seem that a sufficient amount of results are available to draw more general conclusions, it is a most difficult task. The difficulty lies in the fact that the cytochemical differentiation of the individual lipid groups is limited to a considerable extent. Although some cytochemical examinations indicated that mammalian ova especially contained the group of neutral lipids and phospholipids, their chemical verification has only been carried out recently and has been shown by Korolev and Zavarzina (1976) by chromatographic separation of rabbit ova extracts; they also demonstrated the presence of cholesterol. Among phospholipids, they found the representation of cephalin as well as of lecithin and sphingomyelin.

Lipids occur in mammalian ova in two basic morphologic forms. The more conspicuous is the homophasic form — fat droplets — which until now has been studied more frequently. Masked lipids, representing the other form, have been studied only rarely (Trávník, 1976, 1977a).

The discovery made by Korolev (1976) is very important for evaluating the changes in the content of fat droplets during cleavage. On the basis of cytochemical investigations of a number of mammalian species, the author assumes that the amount of lipids depends on the systematic classification. According to the size of ova and the content of fat droplets, the author divides mammals into three groups. In the first group consists at large ova with a high amount of lipids, typical of the cat, dog, and pig. A high lipid content in the ova of carnivora was also observed by Enders (1971) and in the pig by Corner (1963). The first changes in the lipid contents, according to Korolev (1976), occur at the stage of the morula, when the lipid contents are lowered in some cells. In blastocysts, the lipid contents in embryoblast and trophoblast cells differ significantly. The former contain small amounts of lipids, whereas the latter are overfilled with lipids. According to Korolev (1976), large ova containing a small amount of lipids belong to the mammals of the second group. Here he includes the ova of the rabbit, hare, dolphin, marten, and man. The changes in the total content of lipids manifest themselves in these species as late as the blastocyst stage and their nature is similar to that of the ova in the first group. Korolev (1976) included in the third group mammals whose ova are small and contain a negligible quantity of lipid inclusions, which has also been confirmed in the rat and mouse by Enders (1971) and Dvořák and Trávník (1972) and in the hamster by Guraya (1975). The lipid content remains the same throughout cleavage.

On the ultrastructural level, the changes in the contents of homophasic lipids (fat droplets) have until now been studied in detail in the rat ovum. Enders (1971) and Dvořák and Trávník (1972) both state that the contents of fat droplets increase during cleavage. The accuracy of these preliminary estimates has also been proved by the morphometric determination of the volume of fat droplets carried out by Dvořák et al. (1974, 1977) in segmenting rat ova. The authors established that the fat droplet volume in the one-cell ovum is lower than 0.01% of the total cytoplasmic volume. In further ovum development, the volume of fat droplets slowly increases (the maximum established in the blastocyst trophoblast is 0.034% of the cytoplasmic volume). Trávník (1976, 1977a) has also studied masked lipids in segmenting rat ova and found that their amount decreases during cleavage. The increase in the number of fat droplets is interpreted by him to be a result of the interconversion between the masked and homophasic forms of lipids; the total amount of lipids probably remains the same. He further thinks that stored lipids do not play a significant role as a source of energy in the process of rat ovum cleavage.

3.5.3. Proteins

Proteins represent one of the most important cell components determining the molecular organization of the cell. They are contained in the ground cytoplasm, and they constitute and essential part of membranes as well as cell organelles and also a part of enzymes, respiratory pigments, hormones, and secretion products. While the synthesis of proteins, including its morphologic picture, has been intensely studied in recent years (Droz et al., 1973), practically little is known about their degradation, reorganization, or possible stored function. Biochemical evidence that the contents of proteins change, i.e., decrease, during cleavage (on the whole by 25%) has been supplied by Brinster (1967c) in mice.

In mammalian ova generally recognized stored structures of protein character are the so-called lamellar structures, also termed paracrystalline lamellae (Szollosi, 1965b), plaques (Schlafke and Enders, 1967), cytoplasmic nonmembranous lamellae (Weakley, 1968), and crystalloids and fibrous material (Calarco and Brown, 1969).

They originate as early as oogenesis in oocytes of growing follicles and have been demonstrated in the oocytes and/or free tubal ova of the rat (Schlafke and Enders, 1967; Szollosi, 1965b; Weakley, 1968; Dvořák et al., 1970, 1972), rabbit (Szollosi, 1965b), hamster (Szollosi, 1965b; Hadek, 1966; Weakley, 1966, 1967, 1968, 1973), mouse (Weakley, 1968; Calarco and Brown, 1969; Hillman and Tasca, 1969), monkey (Hope, 1965), and man (Wartenberg and Stegner, 1960, Zamboni et al., 1966b). Evidence of the protein character of lamellar structures was brought by Weakley (1967), who found that they were digested by pepsine, and recently also by Dvořák et al. (1975), who used various types of fixation and staining and application of some ultra-cytochemical methods.

The decrease in lamellar structures during the cleavage of rat ova is described by Schlafke and Enders (1967) and Dvořák et al. (1970, 1972). This fact has also been proved on the basis of the study of quantitative changes of the partial volume of these structures which was carried out by Dvořák et al. (1974, 1977) in segmenting rat ova. These authors proved that the greatest volume of lamellar structures is in the one-cell ovum (33.4% of the total cytoplasmic volume). A continuous decrease follows; the lowest values have been found in the trophoblast cells of the late blastocyst (11.3% of the total cytoplasmic volume). The gradual decrease in the partial volume of lamellar structures is simultaneously accompanied by morphologic indications of their decomposition (Dvořák et al., 1972). Even though the facts suggest the probable utilization of lamellar structures during cleavage, the nature of their biochemical mobilization remains obscure.

3.5.4. Conclusion

Mammalian ova cover part of their energetic requirement in the course of cleavage by utilizing the endogenous storage material. Especially glycogen fulfills the role of an important storage material. This function can, however, evidently be performed by proteins and probably also by lipids. The metabolic mobilization of glycogen and its degradation in the course of rat ovum cleavage is manifested not only by a gradual reduction in its amount but particularly by the results of the cytochemical determination of the activities and forms of key enzymes of glycogen metabolism. α-glucan phosphorylase is very active almost for the whole preimlantation period and catalyzes the breakdown of glycogen located in the ground cytoplasm. Morphologic evidence exists that secondary lysosomes (small and large autophagic vacuoles and possibly also multivesicular bodies) assert themselves in the degradation of glycogen in the segmenting rat ovum. The importance of autophagic degradation of glycogen in lysosomes during cleavage has yet to be clarified.

Rat ova, but also those of other mammals, contain lipids essentially in the forms of neutral lipids and phospholipids. Differences among the individual species are evident in the total lipid amount. The changes in the amount of lipids during cleavage have only been described in the ova of some carnivora. Together with the completion of morphologic and cytochemical data, it is essential that attention be turned to the study of the mechanisms of their biochemical mobilization. Storage protein structures in rat ova (but also in those of some other mammals) are the so-called lamellar structures. In the course of cleavage, the amount of lamellar structures decreases, which is often accompanied by morphologic signs of their degradation. The nature of the biochemical mobilization and utilization of lamellar structures has not yet been clarified.

Summary

The segmenting mammalian ovum represents a unique object for the study of the first differentiation processes of the embryo, reflected in the changes in the morphology, biochemistry, physiology, and other properties of the ovum. The basis of this study are more than 2500 rat ova (Rattus norvegicus var. alba) obtained after fertilization and during cleavage at the one-, two-, four-, and eight-cell stages, the morula, and the early and the late blastocysts. The ova flushed from the tuba uterina or the uterus were processed especially for electron microscopy, part of them were subjected to ultracytochemical or cytochemical studies. In particular, the location of enzymes – alkaline and acid phosphatases, nonspecific esterase, cholinesterase, succinate dehydrogenase, α-glucan phosphorylase, glycogen synthetase, and endogenous peroxidase – was studied. Further more, the occurrence of polysaccharides, particularly glycogen, and the glycocalyx were investigated. The ingestion of exogenous proteins (e.g., horseradish peroxidase and microperoxidase) was also studied ultracytochemically. Besides the qualitative analysis of the ultrastructural pattern of changes during the cleavage of the rat ovum, the quantitative analysis of the changes in the occurrence of cytoplasmic structures was also carried out by the morphometric method. Comprehensive information is also supplied concerning the functional aspects of morphologic changes during cleavage with respect to data in the literature.

Morphologic changes during rat ovum cleavage are both quantitative and qualitative in character and appearing in the structure of the nucleus as well as the cytoplasm. In the nucleus, besides some temporary and minor changes in the structure of the nuclear envelope and the arrangement of chromatin and/or further components, the submicroscopic structure of the nucleoli differentiates most.

The dynamic development of the structure of the nucleolus, a high number of compact nucleoli, and clear signs of extrusion of nucleolar material at the early stages of rat ova are typical. A gradual development of compact nucleoli into nucleoli of reticular type occurs during rat ovum cleavage like in all mammalian species studied so far. The reconstruction of nucleoli can be connected to the increasing synthesis of the rRNA, causing the number of ribosomes to grow. The morphology of the nucleolus must, however, be evaluated with respect to the fact that it depends on the phase of the cell cycle.

Conspicuous and numerous changes can be observed in the cytoplasm, both of the cell organelles and cytoplasmic inclusions. Changes in the number, shape, and inner structure of mitochondria occur, and the volume of mitochondria remains substantially unchanged. Correlations of cytochemical, ultracytochemical, and morphologic data show that transformations in fine structure, characterized by a different arrangement, course, and number of cristae, can be related to metabolic processes. Succinate dehydrogenase occurs in the mitochondria of all developmental stages of ova studied.

Lysosomes belong to the most frequently represented cytoplasmic structures in the rat ovum during the whole preimplantation period. Primary lysosomes, in the form of smooth vesicles exhibiting acid phosphatase and nonspecific esterase activity, are particularly numerous in the early stages of cleavage. Secondary lysosomes are represented in the early stages mostly by multivesicular bodies; their number decreases during the course of cleavage. Autophagic vacuoles are present throughout the whole preimplantation development. The activity of both acid phosphatase and nonspecific esterase was found in autophagic vacuoles; multivesicular bodies only exhibited acid phosphatase activity. The incidence of residual bodies during ovum gradually increased. Lysosomes assert themselves in ova in processes of both nutrition and differentiation, whereby a special part is played by cortical granules in the cortical reaction.

The structures of the granular endoplasmic reticulum appear at the eight-cell stage for the first time and increase in number during further development. Acid phosphatase and organophosphate-sensitive esterase activity was demonstrated in its cisternae; at the eight-cell stage, alkaline phosphatase activity was also found.

Smooth endoplasmic reticulum is present in the first stages of ovum cleavage in two forms; large sacs and small tubules or vesicles grouped into large areas. The volume of the structures of smooth endoplasmic reticulum gradually decreases during cleavage and in later stages of development has the form of small, individually distributed vesicles. Acid and alkaline phosphatase and organophosphate-sensitive esterase were found in the structures of smooth enoplasmic reticulum.

The occurrence of ribosomes in the earliest stages of cleavage has so far not been exactly determined; it is, however, doubtlessly very low. From the four-cell stage up to the blastocyst stage, the number of polysomes increases rapidly. A significant rise in the number of polysomes and structures of granular endoplasmic reticulum is a sign of increased proteosynthetic activity. This is also in accordance with the results of biochemical studies in the segmenting ova of different mammalian species.

The Golgi complex occurs in the form of several small fields at the early stages of cleavage, especially in the cortical cytoplasm and later near the nucleus. In the late stages, the Golgi complex is more prominent. Acid phosphatase activity was ascertained in its cisternae and vesicles; in the early stages, alkaline phosphatase activity was also observed.

Significant changes also take place in the content of endogenous storage material, particularly in glycogen. During cleavage, its amount in the ovum gradually decreases. At the same time, the activity and forms of occurrence of the key enzymes of the glycogen metabolism change dynamically. Interesting from the functional point of view is the autophagic degradation of part of the glycogen in secondary lysosomes of the rat ovum.

Lipids as an endogenous source of energy in the rat ovum hardly assert themselves at all; the number of lipid droplets increases during cleavage. Proteins with a probable storage function are represented in the rat ovum by the lamellar structures. At the end of cleavage, their amount diminishes, and signs of their morphologic degradation at the blastocyst stage are regularly found. No details are known about the nature of the biochemical mobilization of lamellar structures.

During the cleavage of the ovum, the cell membrane undergoes changes which are particularly connected to the development of contacts and junctional structures between the blastomeres. The adhesion of blastomeres increases from the eight-cell stage onward by the presence of junctional structures, which are best differentiated from trophoblast cells of the late blastocyst. The formation of zonulae occludentes at the morula stage is a condition for the formation of the blastocyst cavity. In two-cell and four-cell ova, it is evident that the glycocalyx and/or some material of similar character plays an important role in cell adhesion. The zona pellucida also undergoes morphologic changes during cleavage.

The segmenting mammalian ova are not isolated objects independent of the environment. In passing through the tuba uterina and in the uterus before implantation, they can ingest substances of different chemical character, even large molecules, as has been convincingly shown by in vitro experiments. In controlling those processes, both the zona pellucida and especially the cell membrane with the glycocalyx assert themselves. The ingestion of exogenous material can be of importance for the differentiation and metabolic processes of the ovum.

References

Adams, C. E., Hay, M. F., Lutwak-Mann, C.: The action of various agents upon the rabbit embryo. J. Embryol. Exp. Morphol. **9**, 468–491 (1961)

Adams, C. E., Hertig, A. T.: Studies on guinea pig oocytes. I. Electron microscopic observations on the development of cytoplasmic organelles in oocytes of primordial and primary follicles. J. Cell Biol. **21**, 397–427 (1964)

Aitken, R. J., Burton, J., Hawkins, J., Kerr-Wilson, R., Short, R. V., Steven, D. H.: Histological and ultrastructural changes in the blastocyst and reproductive tract of the roe deer, Capreolus capreolus, during delayed implantation. J. Reprod. Fertil. **34**, 481–493 (1973)

Alfert, M.: A cytochemical study of oogenesis and cleavage in the mouse. J. Cell. Comp. Physiol. **36**, 381–409 (1950)

Anderson, E.: The localization of acid phosphatase and the uptake of horseradish peroxidase in the oocyte and follicle cells of mammals. In: Oogenesis. Biggers, J. D., Schuetz, A. W., (Eds.) Baltimore, Butterworth, London: University Park Press, 1972

Anderson, E., Beams, H. W.: Cytological observations on the fine structure of the guinea pig ovary with special reference to the oogonium, primary oocyte and associated follicle cells. J. Ultrastruct. Res. **3**, 432–446 (1960)

Austin, C. R.: The formation, growth, and conjugation of the pronuclei in the rat egg. J. R. Microsc. Soc. **71**, 295–306 (1951)

Austin, C. R.: The development of the pronuclei in the rat egg, with particular references to quantitative relations. Aust. J. Sci. Res. B **5**, 354–365 (1952)

Austin, C. R.: Cortical granules in hamster eggs. Exp. Cell Res. **10**, 533–540 (1956)

Austin, C. R.: The Mammalian Egg. Oxford: Blackwell Scientific Publications, 1961

Austin, C. R., Lovelock, J. E.: Permeability of rabbit, rat and hamster egg membranes. Exp. Cell Res. **15**, 260–261 (1958)

Baca, M., Zamboni, L.: The fine structure of human follicular oocytes. J. Ultrastruct. Res. **19**, 354–381 (1967)

Baeckeland, E.: Glycogen content of rat eggs during cleavage. C. R. Soc. Biol. (Paris) **169**, 452–455 (1975a)

Baeckeland, E.: Mise evidence, teneur et repartition du glycogene dans des oeufs de Rat en segmentation, Bull. Assoc. Anat. **59**, 93–101 (1975b)

Barańska, W., Dorywalski, K., Kujawa, M.: Succinic dehydrogenase activity in fertilized and unfertilized mouse eggs after parthenogenetic stimulation. Folia Histochem. Cytochem. **11**, 303–304 (1973)

Barańska, W., Konwiński, M., Kujawa, M.: Fine structure of the zone pellucida of unfertilized egg cells and embryos. J. Exp. Zool. **192**, 193–202 (1975)

Barlow, P., Owen, D. A. J., Graham, C.: DNA synthesis in the preimplantation mouse embryo. J. Embryol. Exp. Morphol. **27**, 431–445 (1972)

Baudhuin, P., Hers, H. G., Loeb, H.: An electron microscopic and biochemical study of type II glycogenosis. Lab. Invest. **13**, 1139–1152 (1964)

Beier, H. M.: Protein patterns of endometrial secretion in the rabbit. In: Ovo-implantation, Human Gonadotropins and Prolactin. Hubinot, P. O., Leroy, F., Robyn, C., Leleuse, P., (Eds.) Brussels: S. Karger, 1970

Bennett, H. S.: Morphological aspects of extracellular polysaccharides. J. Histochem. Cytochem. **11**, 14–23 (1963)

Bennett, H. S.: The cell surface: components and configurations. In: Handbook of Molecular Cytology. Lima-de-Faria, A., (Ed.). Amsterdam-London: North-Holland Publishing Co., 1969

Biggers, J. D., Stern, S.: Metabolism of the preimplantation mammalian embryo. Adv. Reprod. Physiol. **6**, 2–59 (1973)

Biggers, J. D., Whittingham, D. G., Donahue, R. P.: The pattern of energy metabolism in the mouse oocyte and zygote. Proc. Natl. Acad. Sci. (USA) **58**, 560–567 (1967)

Björkman, N.: A study of the ultrastructure of the granulosa cells of the rat ovary. Acta Anat. (Basel) **51**, 125–147 (1962)

Blanchette, E. J.: A study of the fine structure of the rabbit primary oocyte. J. Ultrastruct. Res. **5**, 349–363 (1961)

Blerkom, J. van, Manes, C.: Development of preimplantation rabbit embryos in vivo and in vitro. II. A comparison of qualitative aspects of protein synthesis. Dev. Biol. **40**, 40–51 (1974)

Boell, E. J., Nicholas, J. S.: Respiratory metabolism of the mammalian egg. J. Exp. Zool. **109**, 267–281 (1948)

Boyer, P. D.: Carboxyl activation as a possible common reaction in substrate-level and oxidative phosphorylation and in muscle contraction. In: Oxidases and Related Redox Systems. King, T. E., Mason, H. S., Morrison, M., (Eds.) New York: John Wiley, Vol. 2, 1965

Braden, A. W. H., Austin, C. R., David, H. A.: The reaction of zona pellucida to sperm penetration. Aust. J. Biol. Sci. **7**, 391–409 (1954)

Bretscher, M. S., Raff, M. C.: Mammalian plasma membranes. Nature (London) **258**, 43–49 (1975)

Brinster, R. L.: Studies on the development of mouse embryos in vitro. II. The effect of energy source. J. Exp. Zool. **158**, 59–68 (1965a)

Brinster, R. L.: Studies on the development of mouse embryos in vitro. IV. Interactions of energy sources. J. Reprod. Fertil. **10**, 227–240 (1965b)

Brinster, R. L.: Lactate dehydrogenase activity in the preimplanted mouse embryo. Biochim. Biophys. Acta **110**, 439–441 (1965c)

Brinster, R. L.: Carbon dioxide production from glucose by the preimplantation mouse embryo. Exp. Cell Res. **47**, 271–277 (1967a)

Brinster, R. L.: Lactate dehydrogenase activity in preimplantation rat embryo. Nature (London) **214**, 1246–1247 (1967b)

Brinster, R. L.: Protein content of the mouse embryo during the first fice days of development. J. Reprod. Fertil. **13**, 413–420 (1967c)

Brinster, R. L.: Carbon dioxide production from glucose by the preimplantation rabbit embryo. Exp. Cell Res. **51**, 330–334 (1968)

Brinster, R. L.: Incorporation of carbon from glucose and pyruvate into the preimplantation mouse embryo. Exp. Cell Res. **58**, 153–158 (1969)

Brinster, R. L.: Culture of two-cell rabbit embryos to morulae. J. Reprod. Fertil. **21**, 17–22 (1970)

Brinster, R. L.: Measuring embryonic enzyme activity. In: Methods of Mammalian Embryology. Daniel, J. C., (Ed.) San Francisco: Freeman, 1971

Brinster, R. L., Thomson, J. L.: Development of eight-cell mouse embryos in vitro. Exp. Cell Res. **42**, 303–315 (1966)

Brinster, R. L., Wiebold, J. L., Brunner, S.: Soluble proteins in preimplantation mouse ova. J. Cell Biol. **67**, 43a (1975)

Busch, H., Smetana, K.: The Nucleolus. New York and London: Academic Press, 1970

Calarco, P. G., Brown, E. H.: An ultrastructural and cytological study of preimplantation development of the mouse. J. Exp. Zool. **171**, 253–284 (1969)

Calarco, P. G., Epstein, Ch. J.: Cell surface changes during preimplantation development in the mouse. Dev. Biol. **32**, 208–213 (1973)

Capaldi, R. A.: A dynamic model of cell membranes. Sci. Am. **230**, No. 3, 27–33 (1974)

Chance, B., Williams, G. R.: The respiratory chain and oxidative phosphorylation. Adv. Enzymol. **17**, 65–134 (1956)

Chang, M. C., Hunt, D. M.: Effects of proteolytic enzymes on the zona pellucida of fertilized and unfertilized mammalian eggs. Exp. Cell Res. **11**, 497–499 (1956)

Chappel, J. B.: Systems for the transport of substances into mitochondria. Br. Med. Bull. **24**, 150 (1966)

Chiquoine, A. D.: The development of the zona pellucida of the mammalian ovum. Am. J. Anat. **106**, 149–170 (1960)

Clegg, K. B., Pikó, L.: Patterns of RNA synthesis in early mouse embryos. J. Cell Biol. **67**, 72a (1975)

Cocucci, S. M., Sussman, M.: RNA in cytoplasmic and nuclear fractions of cellular slime mold amebas. J. Cell Biol. **45**, 399–407 (1970)

Corner, G. W.: Cytology of the ovum, ovary and Fallopian tube. In: Special Cytology. Cowdry, E. V., (Ed.) New York–London: Hafner Publish. Comp. Vol. 3, 1963

Cross, P. C., Brinster, R. L.: The sensitivity of one-cell mouse embryos to pyruvate and lactate. Exp. Cell Res. **77**, 57–62 (1973)

116

Čech, S.: Topochemistry of glycogen, α-glucan phosphorylase and glycogen synthetase during the cleavage of rat ovum. (In Czech). Thesis, J. E. Purkyně Univ. Brno (1977a)

Čech, S.: Ultrahistochemischer Nachweis von Glykogen in den frühesten Entwicklungsstadien der Ratte. Verh. Anat. Ges. 71, 583–586 (1977b)

Čech, S.: Glycogen distribution and its changes in rat ovum during cleavage. Acta Fac. Med. Univ. Brun., in press (1977c)

Čech, S.: Histochemical demonstration of α-glucan phosphorylase in the segmenting rat ovum. (In Czech). Scripta Med. 50, 230–231 (1977d)

Daems, W. Th., Wisse, E., Brederoo, P.: Electron microscopy of the vacuolar apparatus. In: Lysosomes in Biology and Pathology. Dingle, J. T., Fell, H. B., (Eds.). Amsterdam: North-Holland Publishing Co., Vol. 1, 1969

Dalcq, A. M., Pasteels, J.: Détermination photométrique de la teneur relative en DNA des noyaux dans les oeufs en segmentation du rat et de la souris. Exp. Cell Res., Suppl. 3, 72–97 (1955)

Dan, K.: Cyto-embryology of echinoderms and amphibia. Int. Rev. Cytol. 9, 321–367 (1960)

Daniel, J. C., Jr.: The pattern of utilization of respiratory metabolic intermediates by preimplantation rabbit embryos in vitro. Exp. Cell Res. 47, 619–624 (1967)

Daniel, J. C., Jr., Krishnan, R. S.: Studies on the relationship between uterine fluid components and the diapausing state of blastocysts from mammals having delayed implantation. J. Exp. Zool. 172, 267–281 (1969)

Danielli, J. F., Davson, H. A.: A contribution to the theory of permeability of thin films. J. Cell. Comp. Physiol. 5, 495–508 (1935)

Dass, C. M. S., Mohla, S., Prasad, M. R. N.: Time sequence of action of estrogen on nucleic acid and protein synthesis in the uterus and blastocyst during delayed implantation in the rat. Endocrinology 85, 528–537 (1969)

Davidson, E. H.: Gene Activity in Early Development. New York and London: Academic Press, 1969

Defrise, A.: Some observations on the living eggs and blastulae of the albino rat. Anat. Rec. 54, 239–250 (1933)

Denker, H. W.: Topochemie hochmolekularer Kohlenhydratsubstanzen in Frühentwicklung und Implantation des Kaninchens. I. Allgemeine Lokalisierung und Charakterisierung hochmolekularer Kohlenhydratsubstanzen in frühen Embryonalstadien. Zool. Jahrb. Physiol. 75, 141–245 (1970a)

Denker, H. W.: Topochemie hochmolekularer Kohlenhydratsubstanzen in Frühentwicklung und Implantation des Kaninchens. II. Beiträge zu entwicklungsphysiologischen Fragestellungen. Zool. Jahrb. Physiol. 75, 246–308 (1970b)

Denker, H. W.: Enzym-Topochemie von Frühentwicklung und Implantation des Kaninchens. I. Glykogenstoffwechsel. Histochemie 25, 256–267 (1971)

Dickmann, Z., Noyes, R. W.: The zona pellucida at the time of implantation. Fertil. Steril. 12, 310–318 (1961)

Dickson, A. D.: Disappearance of the zona pellucida from the rat blastocyst. J. Anat. 97, 620–621 (1963)

Droz, B., Pisam, M., Chrétien, M.: Morphological aspects of protein synthesis. In: Protein Synthesis in Reproductive Tissue. Diczfalusy, A., (Ed.). Stockholm: Karolinska Institutet, 1973

Ducibella, T., Albertini, D. F., Anderson, E., Biggers, J. D.: The preimplantation mammalian embryo: Characterization of intercellular junctions and their appearance during development. Dev. Biol. 45, 231–250 (1975)

Ducibella, T., Anderson, E.: Cell shape and membrane changes in the eight-cell mouse embryo: prerequisites for morphogenesis of the blastocyst. Dev. Biol. 47, 45–48 (1975)

Du Praw, E. J.: Cell and Molecular Biology. New York and London: Academic Press, 1969

Duve, C. de, Wattiaux, R.: Functions of lysosomes. Ann. Rev. Physiol. 28, 435–492 (1966)

Dvořák, M.: Differentiation of rat ovum during cleavage. In: Submicroscopic Cytodifferentiation. Ergeb. Anat. Entwickl.-Gesch. 45, 10–24 (1971)

Dvořák, M.: Submicroscopic differentiation of nucleolus during cleavage of rat ovum. Scripta Med. 47, 497–502 (1974a)

Dvořák, M.: Origin and development of lysosomes and peroxisomes. In: Biogenesis of Cell Organelles. Dvořák, M., (Ed.) Acta Fac. Med. Univ. Brun. 49, 59–86 (1974b)

Dvořák, M.: Morphological differentiation of the surface coats in the rat ova during cleavage. Acta Fac. Med. Univ. Brun., in press (1977)

Dvořák, M.: Morphological differentiation of cell contacts in rat ova during cleavage. Acta Univ. Carol. Med., in press (1978)

Dvořák, M., Kukletová, M., Šťastná, J.: Lamellar structures of ovarian and segmenting rat ova – their occurrence and ultrastructure. (In Czech). Scripta Med. **43**, 310 (1970)

Dvořák, M., Kukletová, M., Šťastná, J.: Ultrastructure and occurrence of lamellar structures in ovarian and segmenting ova in rat. Scripta Med. **45**, 35–45 (1972)

Dvořák, M., Trávník, P.: Occurrence and location of glycogen and lipid particles in the rat ovum during cleavage. Scripta Med. **45**, 553 (1972)

Dvořák, M., Trávník, P.: The incidence of exogenous peroxidase in rat ovum during cleavage (In Russian). Arch. Anat. Gist. Embryol. **69**, 26–29 (1975)

Dvořák, M., Trávník, P.: Uptake of microperoxidase by segmenting rat ova in vitro. Histochemistry **47**, 257–262 (1976)

Dvořák, M., Trávník, P., Staňková, J.: A quantitative incidence of cytoplasmic structures in rat ovum during cleavage. (In Czech). Scripta Med. **47**, 491–492 (1974)

Dvořák, M., Trávník, P., Staňková, J.: A quantitative analysis of the incidence of certain cytoplasmic structures in the ovum of the rat during cleavage. Cell Tissue Res. **179**, 429–437 (1977)

Dvořák, M., Trávník, P., Staňková, J., Šťastná, J. Čech, S.: Lamellar structures in rat ova and their chemical composition. Z. Mikrosk. Anat. Forsch. **89**, 915–921 (1975)

Dvořák, M., Trávník, P., Šťastná, J.: Occurrence of polysaccharides in the rat ovum during cleavage. Folia Morphol. (Prague) **21**, 150–151 (1973)

Eager, D. D., Johnson, M. H., Thurley, K. W.: Ultrastructural studies on the surface membrane of the mouse egg. J. Cell Sci. **22**, 345–353 (1976)

El-Banna, A., Daniel, J. C., Jr.: The effect of protein fractions from rabbit uterine fluids on embryo growth and uptake of nucleic acid and protein precursors. Fertil. Steril. **23**, 105–114 (1972)

Ellem, K. A. O., Gwatkin, R. B.-L.: Patterns of nucleic acid synthesis in the early mouse embryo. Dev. Biol. **18**, 311–330 (1968)

Enders, A. C.: The structure of the armadillo blastocyst. J. Anat. **96**, 39–48 (1962)

Enders, A. C.: The fine structure of the blastocyst. In: The Biology of the Blastocyst. Blandau, R. J., (Ed.) Chicago and London: The University of Chicago Press, 1971

Enders, A. C., Schlafke, S.: A morphological analysis of the early implantation stages in the rat. Am. J. Anat. **120**, 185–226 (1967)

Enders, A. C.: Schlafke, S.: Surface coats of the mouse blastocyst and uterus during the preimplantation period. Anat. Rec. **180**, 31–46 (1974)

Engel, W., Kreutz, R.: Lactate dehydrogenase isoenzymes in the mammalian egg: investigation by micro disc electrophoresis in 15 species of the orders rodentia, lagomorpha, carnivora, artiodactyla and in man. Humangenetik **19**, 253–260 (1973)

Engel, W., Petzold, U.: Early developmental changes of the lactate dehydrogenase isozyme pattern in mouse, rat, guinea pig. Syrian hamster and rabbit. Humangenetik **20**, 125–131 (1973)

Eppig, J. J.: Analysis of mouse oogenesis in vitro. Oocyte isolation and the utilization of exogenous energy. Sources by growing oocytes. J. Exp. Zool. **198**, 375–382 (1976)

Epstein, Ch. J.: Gene expression and macromolecular synthesis during preimplantation embryonic development. Biol. Reprod. **12**, 82–105 (1975)

Epstein, Ch. J., Kwok, L., Smith, S.: The source of lactate dehydrogenase in preimplantation mouse embryos. FEBS Lett. **13**, 45–48 (1971)

Epstein, Ch. J., Smith, S. A.: Amino acid uptake and protein synthesis in preimplantation mouse embryos. Dev. Biol. **33**, 171–184 (1973)

Epstein, Ch. J., Smith, S. A.: Electrophoretic analysis of proteins synthesized by preimplantation mouse embryos. Dev. Biol. **40**, 233–244 (1974)

Ericsson, J. L. E.: Mechanism of cellular autophagy. In: Lysosomes in Biology and Pathology, Dingle, J. T., Fell, H. B., (Eds.) Amsterdam and London: North-Holland Publishing Co., Vol. 2, 1969

Ericsson, J. L. E., Trump, B. F.: Observations on the application of electron microscopy of the lead phosphate technique for the demonstration of acid phosphatase. Histochemie **4**, 470–487 (1964–65).

Farquhar, M. G., Palade, G. E.: Junctional complexes in various epithelia. J. Cell Biol. **17**, 375–412 (1963)

118

Ferm, V. H.: Permeability of the mammalian blastocyst to teratogens. In: The Biology of the Blastocyst. Blandau, R. J., (Ed.) Chicago and London: The University of Chicago Press, 1971

Fléchon, J.-E.: Nature glycoprotéique des granules corticaux de l'oeuf de lapine. Mise en évidence par l'utilisation comparée de techniques cytochimiques ultrastructurales. J. Microsc. 9, 221–242 (1970)

Franchi, L. L., Mandl, A.: The ultrastructure of oogonia and oocytes in foetal and neonatal rats. Proc. R. Soc. Lond. (Biol.) 157, 99–114 (1962)

Fridhandler, L.: Pathways of glucose metabolism in fertilized rabbit ova at various pre-implantation stages. Exp. Cell Res. 22, 303–316 (1961)

Fridhandler, L., Hafez, E. S. E., Pincus, G.: Developmental changes in the respiratory activity of rabbit ova. Exp. Cell Res. 13, 132–139 (1957)

Ghosh, S.: The nucleolar structure. Int. Rev. Cytol. 44, 1–28 (1976)

Gianguzza, M., Mulnard, J.: Quelques aspects ultrastructuraux de l'activité lysosomiale dans les premiers stades du développement de la souris. Arch. Biol. (Liège) 83, 499–512 (1972)

Gibson, C., Masters, C. J.: On the lactate dehydrogenase of preimplantation mouse ova. FEBS Lett. 7, 277 (1970)

Ginsberg, L., Hillman, N.: ATP metabolism in cleavage-staged mouse embryos. J. Embryol. Exp. Morphol. 30, 267–282 (1973)

Glass, L.E.: Transfer of native and foreign serum antigens to oviductal mouse eggs. Am. Zool. 3, 135–156 (1963)

Glass, L. E., McClure, T. R.: Postnatal development of the mouse oviduct: transfer of serum antigens to the tubal epithelium. In: Preimplantation Stages of Pregnancy. Wolstenholme, G. E. W., O'Connor, M., (Eds.) London: J. & A. Churchill, 1965

Golbus, M. S., Calarco, P. G., Epstein, Ch. J.: The effects of inhibitors of RNA synthesis (α-amanitin and actinomycin D) on preimplantation mouse embryogenesis. J. Exp. Zool. 186, 207–216 (1973)

Gonzáles, S. P., Nardone, R. M.: Cyclic nucleolar changes during the cell cycle. I. Variations in number, size, morphology and position. Exp. Cell Res. 50, 599–615 (1968)

Gordon, G. B., Miller, L. R., Bensch, K. G.: Studies on the intracellular digestive process in mammalian tissue culture cells. J. Cell Biol. 25, 41–55 (1965)

Gordon, M., Fraser, L. R., Dandekar, P. V.: The effect of ruthenium red and concanavalin A on the vitelline surface of fertilized and unfertilized rabbit ova. Anat. Rec. 181, 95–112 (1975)

Graham, R. C., Karnovsky, M. J.: The early stages of absorption of injected horseradish peroxidase in the proximal tubules of the mouse kidney: Ultrastructural cytochemistry by a new technique. J. Histochem. Cytochem. 14, 291–302 (1966)

Grasso, J. A., Woodard, J. W.: The relationship between RNA synthesis and hemoglobin synthesis in amphibian erythropoiesis. J. Cell Biol. 31, 279–294 (1966)

Graves, C. N., Biggers, J. D.: Carbon dioxide fixation by preimplantation mouse embryos. Science 167, 1506–1507 (1970)

Gulyas, B. J.: Nuclear extrusion in rabbit embryos. Z. Zellforsch. 120, 151–159 (1971)

Gulyas, B. J., Daniel, J. C.: Oxygen consumption in diapausing blastocyst. J. Cell. Comp. Physiol. 70, 33–36 (1967)

Gulyas, B. J., Krishnan, R. S.: Current status of the chemistry and biology of "blastokinin". In: The Biology of the Blastocyst. Blandau, R. J., (Ed.) Chicago and London: The University of Chicago Press, 1971

Guraya, S. S.: Histochemical nature of cortical granules in the human egg. Z. Zellforsch. 94, 32–35 (1969)

Guraya, S. S.: Histochemical observation of the juxtanuclear complex of organelles in the hamster oocyte. Acta Anat. (Basel) 93, 325–343 (1975)

Gwatkin, R. B. L.: Effect of viruses on early mammalian development: III. Further studies concerning the interaction of Mengo encephalitis virus with mouse ova. Fertil. Steril. 17, 411–420 (1966)

Gwatkin, R. B. L.: Passage of mengovirus through the zona pellucida of the mouse morula. J. Reprod. Fertil. 13, 577–578 (1967)

Gwatkin, R. B. L.: Fertilization. In: The Cell Surface in Animal Embryogenesis and Development. Poste, G., Nicolson, G. L., (Eds.) Amsterdam, New York and London: North-Holland Publishing Co., 1976

119

Gwatkin, R. B. L., Williams, D. T., Hartmann, J. F., Kniazuk, M.: The zona reaction of hamster and mouse eggs: production in vitro by a trypsin-like protease from cortical granules. J. Reprod. Fertil. **32**, 259–265 (1973)

Hackenbrock, C. R.: Ultrastructural bases for metabolically linked mechanical activity in mitochondria. I. Reversible ultrastructural changes with change in metabolic steady state in isolated liver mitochondria. J. Cell Biol. **30**, 269–297 (1966)

Hackenbrock, C. R.: Ultrastructural bases for metabolically linked mechanical activity in mitochondria. II. Electron transport – linked ultrastructural transformations in mitochondria. J. Cell Biol. **37**, 345–369 (1968)

Haddad, A., Nagai, M. E. T.: Radioautographic study of glycoprotein biosynthesis and renewal in the ovarian follicles of mice and the origin of the zona pellucida. Cell Tissue Res. **177**, 347–369 (1977)

Hadek, R.: Submicroscopic study on the sperm-induced cortical reaction in the rabbit ovum. J. Ultrastruct. Res. **9**, 99–109 (1963)

Hadek, R.: Cytoplasmic whorls in the golden hamster oocyte. J. Cell Sci. **1**, 281–282 (1966)

Hadek, R.: Mammalian Fertilization. An Atlas of Ultrastructure. New York and London: Academic Press, 1969

Hamana, K., Hafez, E. S. E.: Disc electrophoretic patterns of uteroglobin and serum proteins in rabbit blastocoelic fluid. J. Reprod. Fertil. **21**, 555–558 (1970)

Hamburgh, M.: Theories of Differentiation. London: Edward Arnold Ltd., 1971

Hanker, J. S., Yates, P. E., Clapp, D. H., Anderson, W. A.: New methods for the demonstration of lysosomal hydrolases by the formation of osmium blacks. Histochemie **30**, 201–214 (1972)

Harris, R. A., Penniston, J. T., Asai, J., Green, D. E.: The conformational basis of energy transformations in membrane systems. II. Correlation between conformational change and functional states. Proc. Natl. Acad. Sci. (USA) **59**, 830 (1968)

Hartmann, J. F., Gwatkin, R. B. L.: Alteration of sites on the mammalian sperm surface following capacitation. Nature (London) **234**, 479–481 (1971)

Hastings, R. A.: Exogenous protein uptake by rabbit preimplantation stages. Anat. Rec. **175**, 339 (1973)

Hastings, R. A. II, Enders, A. C.: Uptake of exogenous protein by the preimplantation rabbit. Anat. Rec. **179**, 311–330 (1974a)

Hastings, R. A. II, Enders, A. C.: Junctional complexes in the preimplantation rabbit embryo. Anat. Rec. **181**, 17–34 (1974b)

Hastings, R. A. II, Enders, A. C., Schlafke, S.: Permeability of the zona pellucida to protein tracers. Biol. Reprod. **7**, 288–296 (1972)

Heldt, H. W.: Energiestoffwechsel in Mitochondrien. Angew. Chemie **84**, 792–797 (1972)

Hers, H. G.: Alpha-glucosidase deficiency in generalized glycogen storage disease (Pompe's disease). Biochem. J. **86**, 11–16 (1963)

Hers, H. G., Wulf, H. de, Stalmans, W.: The control of glycogen metabolism in the liver. FEBS Lett. **12**, 73–82 (1970)

Hertig, A. T., Adams, E. C.: Studies on the human oocyte and its follicle. I. Ultrastructural and histochemical observations on the primordial follicle stage. J. Cell Biol. **34**, 647–675 (1967)

Hesseldahl, H.: Ultrastructure of early cleavage stages and preimplantation in the rabbit. Z. Anat. Entwickl. Gesch. **135**, 139–155 (1971)

Heyner, S., Brinster, R. L., Palm, J.: Effect of iso-antibody on preimplantation mouse embryos. Nature (London) **222**, 783–784 (1969)

Hillman, N., Tasca, R.: Ultrastructural and autoradiographic studies of mouse cleavage stages. Am. J. Anat. **126**, 151–174 (1969)

Hillman, N. W., Tasca, R. J., Wileman, G.: Ultrastructural studies of preimplantation mouse embryos. J. Cell Biol. **35**, 56A (1967)

Holmes, P. V., Dickson, A. D.: Estrogen-induced surface coat and enzyme changes in the implanting mouse blastocyst. J. Embryol. Exp. Morphol. **29**, 639–645 (1973)

Holmes, P. V., Dickson, A. D.: Temporal and spatial aspects of oestrogen-induced RNA, protein and DNA synthesis in delayed-implantation mouse blastocysts. J. Anat. **119**, 453–459 (1975)

Holtzman, E.: Lysosomes in the physiology and pathology of neurons. In: Lysosomes in Biology and Pathology. Dingle, J. T., Fell, H. B., (Eds.) Amsterdam: North-Holland Publishing Co., Vol. I, (1969)

Holtzman, E.: Cytochemical studies of protein transport in the nervous system. Philos. Trans. R. Soc. Lond. (Biol.) **261**, 407–421 (1971)

Holtzman, E.: Lysosomes: A survey. Wien, New York: Springer Verlag 1976

Hope, J.: The fine structure of the developing follicle of the rhesus ovary. J. Ultrastruct. Res. **12**, 592–610 (1965)

Hsu, Y.-C.: Differentiation in vitro of mouse embryos to the stage of early somite. Dev. Biol. **33**, 403–411 (1973)

Hsu, Y.-C., Baskar, J., Stevens, L. C., Rash, J. E.: Development in vitro of mouse embryos from the two-cell egg stage to the early somite stage. J. Embryol. Exp. Morphol. **31**, 235–245 (1974)

Huijing, F.: Glycogen metabolism and glycogen-storage diseases. Physiol. Rev. **55**, 609–658 (1975)

Inoue, M.: Nucleic acids and protein synthesis in the rat embryos in the uterus in the preimplantation stages. Acta Obstet. Gynaecol. Jpn. **18**, 251–260 (1971)

Inoue, M., Wolf, D. P.: Sperm binding characteristics of the murine zona pellucida. Biol. Reprod. **13**, 340–346 (1975)

Ishida, K.: Cytochemical studies of tubal ova. Tohoku J. Agric. Res. **5**, 1–11 (1954)

Ishida, K.: Histochemical studies of the tubal ova of rabbits and pigs. Jpn. J. Fertil. Steril. **8**, 11–17 (1963)

Ishida, K.: Histochemical demonstration of phosphorylase and UDPG-glycogen transferase in fertilized and unfertilized mammalian eggs. Arch. Histol. Jpn. **29**, 447–457 (1968)

Ishida, K., Chang, M. C.: Histochemical demonstration of succinic dehydrogenase in hamster and rabbit eggs. J. Histochem. Cytochem. **13**, 470–475 (1965)

Ito, S.: Structure and function of the glycocalyx. Fed. Proc. **28**, 12–25 (1969)

Izquierdo, L., Vial, J. D.: Electron microscope observations on the early development of the rat. Z. Zellforsch. **56**, 157–179 (1962)

Jacobson, M. A., Sanyal, M. K., Meyer, R. K.: Effect of estrone on RNA synthesis in preimplantation blastocyst of gonadotrophin-treated immature rats. Endocrinology **86**, 982–987 (1970)

Jézéquel, A., Arakawa, K., Steiner, J. W.: The fine structure of the normal, neonatal mouse liver. Lab. Invest. **14**, 1894–1930 (1965)

Jones, L. T., Heap, R. B., Perry, J. S.: Protein synthesis in vitro by pig blastocyst tissue before attachment. J. Reprod. Fertil. **47**, 129–131 (1976)

Kang, Y.-H.: Development of the zona pellucida in the rat oocyte. Am. J. Anat. **139**, 535–566 (1974)

Kang, Y.-H., Anderson, W. A.: Ultrastructure of the oocytes of the egyptian spirty mouse (Acomys cahirinus). Anat. Rec. **182**, 175–200 (1975)

Karnovsky, M. J.: The localization of cholinesterase activity in rat cardiac muscle by electron microscopy. J. Cell Biol. **23**, 217–232 (1964)

Karp, G. C., Manes, C., Hahn, W. E.: Ribosome production and protein synthesis in the preimplantation rabbit embryo. Differentiation **2**, 65–73 (1974)

Keberle, H., Faigle, J. W., Fritz, H., Knusel, F., Loustalot, P., Schmid, K.: Theories on the mechanism of action of thalidomide. In: Embryopathic Activity of Drugs. Robson, J. M., Sullivan, F. M., Smith, R. L., (Eds.) Boston: Little, Brown and Co., 1965

Knowland, J., Graham, C.: RNA synthesis at the two-cell stage of mouse development. J. Embryol. Exp. Morphol. **27**, 167–176 (1972)

Korb, J.: Mitochondrial DNA. (In Czech). Biol. Listy **41**, 124–127 (1976)

Korolev, V. A.: Morpho-functional characteristics of lipids in early embryogenesis of placentale mammals. (In Russian). Arch. Anat. Gist. Embryol. **70**, 18–26 (1976)

Korolev, V. A., Zavarzina, G. A.: Lipids in oocytes of the placental mammals. (In Russian). Citologija **18**, 1281–1284 (1976)

Kotoulas, O. B., Ho, J., Adachi, F., Weigensberg, B. I., Phillips, M. J.: Fine structural aspects of the mobilization of hepatic glycogen. II. Inhibition of glycogen breakdown. Am. J. Pathol. **63**, 23–36 (1971)

Kotoulas, O. B., Phillips, M. J.: Fine structural aspects of the mobilization of hepatic glycogen. I. Acceleration of glycogen breakdown. Am. J. Pathol. **63**, 1–22 (1971)

Kramen, M., Biggers, J. D.: Uptake of tricarboxylic acid cycle intermediates by preimplantation mouse embryos in vitro. Proc. Natl. Acad. Sci. (USA) **68**, 2556–2659 (1971)

Krauskopf, Ch.: Elektronenmikroskopische Untersuchungen über die Struktur der Oozyte und des 2-Zellenstadiums beim Kaninchen. II. Blastomeren. Z. Zellforsch. **92**, 296–312 (1968)

121

Krishnan, R. S., Daniel, J. C., Jr.: "Blastokinin": Inducer and regulator of blastocyst development in the rabbit uterus. Science 158, 490–492 (1967)

Kukletová, M., Lukáš, Z., Mazanec, K.: Submicroscopic localization of nonspecific esterase in the oocyte and 2-cell stage of the rat ovum. Z. Mikrosk. Anat. Forsch. 88, 455–464 (1974)

McLaren, A.: The fate of the zona pellucida in mice. J. Embryol. Exp. Morphol. 23, 1–19 (1970)

Lehninger, A. L.: The Mitochondrion: Molecular Basis of Structure and Function. New York: W. A. Benjamin Inc., 1964

Lehninger, A. L.: Biochemistry. The Molecular Basis of Cell Structure and Function. New York: Worth Publishers, Inc., 1972

Loewenstein, J. E., Cohen, A. I.: Dry mass, lipid content and protein content of the intact and zona-free mouse ovum. J. Embryol. Exp. Morphol. 12, 113–121 (1964)

Longo, F. J.: An ultrastructural analysis of spontaneous activation of hamster eggs aged in vivo. Anat. Rec. 179, 27–56 (1974)

Luft, J. H.: Ruthenium red and violet. I. Chemistry, purification, methods of use for electron microscopy and mechanism of action. Anat. Rec. 171, 347–368 (1971)

Lutwak-Mann, C.: The rabbit blastocyst and its environment: physiological and biochemical aspects. In: The Biology of the Blastocyst. Blandau, R. J., (Ed.) Chicago and London: The University of Chicago Press, 1971

Man, J. C. H. de, Noorduyn, N. J. A.: Ribosomes: Properties and function. In: Handbook of Molecular Cytology. Lima-de-Faria, A., (Ed.) Amsterdam-London: North-Holland Publishing Co., 1969

Manes, C.: Nucleic acid synthesis in preimplantation rabbit embryos. I. Quantitative aspects, relationship to early morphogenesis and protein synthesis. J. Exp. Zool. 172, 303–310 (1969)

Manes, C.: Nucleic acid synthesis in preimplantation rabbit embryos. II. Delayed synthesis of ribosomal RNA. J. Exp. Zool. 176, 87–96 (1971)

Manes, C., Daniel, J. C., Jr.: Quantitative and qualitative aspects of protein synthesis in the preimplantation rabbit embryo. Exp. Cell Res. 55, 261–268 (1969)

Maraldi, N. M., Monesi, V.: Ultrastructural changes from fertilization to blastulation in the mouse. Arch. Anat. Microsc. 59, 361–382 (1970)

Marston, J. H., Chang, M. C.: The morphology and timing of fertilization and early cleavage in the Mongolian gerbil and Deer mouse. J. Embryol. Exp. Morphol. 15, 169–191 (1966)

Martínez-Palomo, A.: The surface coats of animal cells. Int. Rev. Cytol. 29, 29–75 (1970)

Mayahara, H., Hirano, H., Saito, T., Ogawa, K.: The new lead citrate method for the ultracytochemical demonstration of activity of non-specific alkaline phosphatase (orthophosphoric monoester phosphohydrolase). Histochemie 11, 88–96 (1967)

Mazanec, K.: Submikroskopische Veränderungen während der Furchung eines Säugetiereies. Arch. Biol. (Liège) 76, 49–85 (1965)

Mazanec, K., Dvořák, M.: On the submicroscopical changes of the segmenting ovum in the albino rat. Čs. Morfol. 11, 103–108 (1963)

McReynolds, H. D., Hadek, R.: A comparison of the fine structure of late mouse blastocyst developed in vivo and in vitro. J. Exp. Zool. 182, 95–118 (1972a)

McReynolds, H. D., Hadek, R.: Periodic acid Schiff-positive material in hamster preimplantation embryos. J. Reprod. Fertil. 30, 173–175 (1972b)

Menke, T., McLaren, A.: Mouse blastocysts grown in vivo and in vitro: carbon dioxide production and trophoblast outgrowth. J. Reprod. Fertil. 23, 117–127 (1970)

Merchant, H.: Ultrastructural changes in preimplantation rabbit embryos. Cytologia 35, 319–334 (1970)

Mills, R. M., Jr., Brinster, R. L.: Oxygen consumption of preimlantation mouse embryos. Exp. Cell Res. 47, 337–344 (1967)

Mintz, B.: Synthetic processes and early development in the mammalian egg. J. Exp. Zool. 157, 85–100 (1964)

Mitchell, P.: A chemiosmotic hypothesis for the mechanism of oxidative and photosynthetic phosphorylation. Nature (London) 191, 144–148 (1961)

Mitchell, P.: Chemiosmotic coupling in oxidative and photosynthetic phosphorylation. Biol. Rev. 41, 445 (1965)

Mohla, S., Prasad, M. R. N.: Early action of estrogen on the incorporation of (^{3}H)uridine in the blastocyst and uterus of rat during delayed implantation. J. Endocrinol. 49, 87–92 (1971)

Monesi, V., Molinaro, M., Spalletta, E., Davoli, C.: Effect of metabolic inhibitors on macromolecular synthesis and early development in the mouse embryo. Exp. Cell Res. 59, 197–206 (1970)

Monesi, V., Salfi, V.: Macromolecular syntheses during early development in the mouse embryo. Exp. Cell Res. 46, 632–635 (1967)

Mulnard, J., Dalcq, A. M.: Les polysaccharides dans le dévelopment de l'oeuf tubaire du rat. C. R. Soc. Biol. (Paris) 149, 836–839 (1955)

Nadijcka, M., Hillman, N.: Ultrastructural studies of the mouse blastocyst substages. J. Embryol. Exp. Morphol. 32, 675–695 (1974)

Nanninga, N.: Structural aspects of ribosomes. Int. Rev. Cytol. 35, 135–188 (1973)

Norberg, H. S.: The follicular oocyte and its granulosa cells in domestic pig. Z. Zellforsch. 131, 497–517 (1972)

Norberg, H. S.: Ultrastructural aspects of the preattached pig embryo: cleavage and early blastocysts stages. Z. Anat. Entwickl. Gesch. 143, 95–114 (1973a)

Norberg, H. S.: Ultrastructure of pig tubal ova. The unfertilized and pronuclear stage. Z. Zellforsch. 141, 103–122 (1973b)

Norrevang, A.: Electron microscopic morphology of oogenesis. Int. Rec. Cytol. 23, 113–186 (1968)

Odor, L. D.: Electron microscopic studies on ovarian oocytes and unfertilized tubal ova in the rat. J. Biophys. Biochem. Cytol. 7, 567–574 (1960)

Odor, L. D.: The ultrastructure of unilaminar follicles of the hamster ovary. Am. J. Anat. 116, 493–522 (1965)

Ogawa, K., Saito, T., Mayahara, H.: The sites of ferricyanide reduction by reductases within mitochondria as studied by electron microscopy. J. Histochem. Cytochem. 16, 49–57 (1968)

Olds, P. J., Stern, S., Biggers, J. D.: Chemical estimates of the RNA and DNA contents of the early mouse embryo. J. Exp. Zool. 186, 39 (1973)

Ozias, C. B., Weitlauf, H. M.: Hormonal influences on the glycogen content of normal and delayed implanting mouse blastocysts. J. Exp. Zool. 177, 147–152 (1971)

Ozias, C. B., Stern, S.: Glycogen levels of preimplantation mouse embryos developing in vitro. Biol. Reprod. 8, 467–472 (1973)

Panigel, M., Kraemer, D. C., Kalter, S. S., Smith, G. C., Heberling, R. L.: Ultrastructure of cleavage stages and preimplantation embryos of the baboon. Anat. Embryol. 147, 45–62 (1975)

Parkening, T. A., Soderwall, A. L.: Preimplantation stages from young and senescent golden hamsters: presence of succinic dehydrogenase and non-viable ova. J. Reprod. Fertil. 35, 373–376 (1973)

Parkening, T. A., Soderwall, A. L.: Histochemical localization of glycogen in preimplantation and implantation stages of young and senescent golden hamsters. J. Reprod. Fertil. 41, 285–295 (1974)

Petzoldt, U.: Autoradiographic studies on the origin of rabbit blastocyst fluid proteins. Cytobiologie 9, 401–406 (1974a)

Petzoldt, U.: Micro-disc electrophoresis of soluble proteins in rabbit blastocysts. J. Embryol. Exp. Morphol. 31, 479–487 (1974b)

Phillips, M. J., Unakar, N. J., Doornewaard, G., Steiner, J. W.: Glycogen depletion in the newborn rat liver. An electron microscopic and electron histochemical study. J. Ultrastruct. Res. 18, 142–165 (1967)

Pickworth, S., Yerganian, G., Chang, M. C.: Fertilization and early development of the chinese hamster, Cricetulus griseus. Anat. Rec. 162, 197–207 (1968)

Pike, I. L., Murdoch, R. N., Wales, R. G.: The incorporation of carbon dioxide into the major classes of RNA during culture of the preimplantation mouse embryo. J. Reprod. Fertil. 45, 211–226 (1975)

Pike, I. L., Wales, R. G.: The effect of exogenous substrate on the metabolism of glycogen during early development of the preimplantation mouse embryo. J. Reprod. Fertil. 43, 388–389 (1975)

Pikó, L.: Synthesis of macromolecules in early mouse embryos cultured in vitro: RNA, DNA, and a polysaccharide component. Dev. Biol. 21, 257–279 (1970)

Pikó, L., Chase, D. G.: Role of the mitochondrial genome during early development in mice. J. Cell Biol. 58, 357–378 (1973)

Pikó, L., Matsumoto, L.: Number of mitochondria and some properties of mitochondrial DNA in the mouse egg. Dev. Biol. **49**, 1–10 (1976)

Potts, D. M., Racey, P. A.: A light and electron microscope study of early development in the bat Pipistrellus pipistrellus. Micron **2**, 322–348 (1971)

Poznakhirkina, N. A., Serov, O. L., Korochin, L. I.: A study on lactate dehydrogenase isozymes in rat ova. Biochem. Genet. **13**, 65–72 (1975)

Prasad, M. R. N., Dass, C. M., Mohla, S.: Action of esterogen on the blastocyst and uterus in the delayed implantation – an autoradiographic study. J. Reprod. Fertil. **16**, 97–104 (1968)

Prasad, M. R. N., Dass, C. M., Mohla, S.: Time sequence of action of estrogen on nucleic acid and protein synthesis in the uterus and blastocyst during delayed implantation in the rat. Endocrinology **85**, 528–536 (1969)

Procházka, J., Mazanec, K.: Über die Veränderungen der Mitochondrien im Verlauf der Furchung des Ratteneies. Scripta Med. **38**, 201–210 (1965)

Quinn, P., Wales, R. G.: Adenosine triphosphate content of preimplantation mouse embryos. J. Reprod. Fertil. **25**, 133–135 (1971)

Rambourg, A.: Morphological and histological aspects of glycoproteins at the surface of animal cells. Int. Rev. Cytol. **31**, 57–114 (1971)

Reinius, S.: Ultrastructure of blastocyst attachment in the mouse. Z. Zellforsch. **77**, 257–266 (1967)

Rhodin, J. A. G.: Histology, A Text and Atlas. New York-London-Toronto: Oxford University Press, 1974

Robertson, J. D.: The ultrastructure of cell membranes and their derivates. Biochem. Soc. Symp. **16**, 3–43 (1959)

Robertson, J. D.: The molecular structure and contact relationships of cell membranes. Progr. Biophys. **10**, 343–418 (1960)

Rosenfeld, E. I.: Animal tissue γ-amylase and its role in the metabolism of glycogen. In: Control of Glycogen Metabolism. Ciba Foundation Symposium. Whelan, W. J., Cameron, M. P., (Eds.) London: J. & A. Churchill, Ltd., 1964

Rosenfeld, M. G., O'Malley, B. W.: Steroid hormones: effects on adenyl cyclase activity and adenosine 3', 5'-monophosphate in target tissues. Science **168**, 253–255 (1970)

Sanyal, M. K., Myer, R. K.: Effect of esterone on DNA synthesis in preimplantation blastocysts of gonadotrophin-treated immature rats. Endocrinology **86**, 976–981 (1970)

Schiffner, J., Spielmann, H.: Fluorometric assay of the protein content of mouse and rat embryos during preimplantation development. J. Reprod. Fertil. **47**, 145–147 (1976)

Schlafke, S., Enders, A. C.: Cytological changes during cleavage and blastocyst formation in the rat. J. Anat. **102**, 13–32 (1967)

Schlafke, S., Enders, A. C.: Protein uptake by rat preimlantation stages. Anat. Rec. **175**, 539–560 (1973)

Schnaitman, C., Greenawalt, J. W.: Enzymatic properties of the inner and outer membranes of rat liver mitochondria. J. Cell Biol. **38**, 158–175 (1968)

Schuchner, E. B.: Ultrastructural changes of the nucleoli during early development of fertilized rat eggs. Biol. Reprod. **3**, 265–274 (1970)

Schultz, G. A.: Characterization of polyribosomes containing newly synthesized messenger RNA in preimplantation rabbit embryos. Exp. Cell Res. **82**, 168–174 (1973)

Schultz, G., Manes, C., Hahn, W. E.: Synthesis of RNA containing polyadenylic acid sequences in preimplantation rabbit embryos. Dev. Biol. **30**, 418–426 (1973)

Selman, K.: Cortical granules in the golden hamster: their formation, release, and cytochemical characterization. J. Cell Biol. **63**, 309a (1974)

Selman, K., Anderson, E.: The formation and cytochemical characterization of cortical granules in ovarian oocytes of the golden hamster (Mesocricetus auratus). J. Morphol. **147**, 251–274 (1975)

Sidebottom, E., Deák, I.I.: The function of the nucleolus in the expression of genetic information: Studies with hybrid animal cells. Int. Rev. Cytol. **44**, 29–53 (1976)

Siracusa, G.: RNA polymerase during early development in mouse embryo. Exp. Cell Res. **78**, 460–462 (1973)

Sjöstrand, F. S.: The ultrastructure of cells as revealed by the electron microscope. Int. Rev. Cytol. **5**, 455–533 (1956)

124

Skalko, R. G., Morse, J. M. D.: The differential response of the early mouse embryo to actinomy-cin D treatment in vitro. Teratology 2, 47–54 (1969)

Slater, E. C.: Mechanism of phosphorylation in the respiratory chain. Nature (London) 172, 975–982 (1953)

Slater, E. C.: The coupling between energy-yielding and energy-utilizing reactions in mitochondria. Q. Rev. Biophys. 4, 35–71 (1971)

Smith, D. M.: The effect on implantation of treating cultured mouse blastocysts with oestrogen in vitro and the uptake of (^{3}H) oestradiol by blastocysts. J. Endocrinol. 41, 17–29 (1968)

Snyder, T. E., Weitlauf, H. M., Nelson, S. R.: Comparison of the glycogen content of eggs in the uteri and oviducts of intact and hypophysectomized mice. Biol. Reprod. 5, 314–318 (1971)

Solter, D., Damjanov, I., Škreb, N.: Demonstrability of some oxidative enzymes in early rodent embryos with and without fixation. Dev. Biol. 29, 486–490 (1972)

Sorensen, R. A.: Problems in Oocyte Maturation and Early Development in the Mouse. Thesis, Yale University, 1972

Sotelo, J. R., Porter, K. R.: An electron microscope study of the rat ovum. J. Biophys. Biochem. Cytol. 5, 327–342 (1959)

Spielmann, H.: Different patterns of energy metabolism in the rat and mouse zygote. J. Reprod. Fertil. 42, 391–394 (1975)

Spielmann, H., Erickson, R. P., Epstein, C. J.: Immunochemical studies of lactate dehydrogenase and glucose-6-phosphate dehydrogenase in preimplantation mouse embryos. J. Reprod. Fertil. 40, 367–373 (1974)

Starck, D.: Embryologie. Ein Lehrbuch auf allgemein biologischer Grundlage. Stuttgart: Georg Thieme Verlag, 1975

Stegner, H.-E.: Die elektronenmikroskopische Struktur der Eizelle. Ergeb. Anat. Entwickl. Gesch. 39, 1–113 (1967)

Stegner, H.-E., Wartenberg, H.: Elektronenmikroskopische und histotopochemische Untersuchun-gen über Struktur und Bildung der Zona pellucida menschlicher Eizellen. Z. Zellforsch. 53, 702–712 (1961)

Stern, S.: The activity of glycogen synthetase in the cleaving mouse embryo. 3rd Annual Meeting of the Soc. for the Study of Reproduction, Columbus, Ohio, 1970

Stern, S., Biggers, J. D.: Enzymatic estimation of glycogen in the cleaving mouse embryo. J. Exp. Zool. 168, 61–66 (1968)

Stern, S., Biggers, J. D., Anderson, E.: Mitochondria and early development of the mouse. J. Exp. Zool. 176, 179–192 (1971)

Stoeckenius, W., Engelman, D. M.: Current models for the structure of biological membranes. J. Cell Biol. 42, 613–646 (1969)

Stone, L. S., Hammer, C. E.: Biochemistry and physiology of oviductal secretions. Gynecol. Invest. 6, 234–252 (1975)

Straus, W.: Lysosomes, phagosomes, and related particles. In: Enzyme Cytology. Roodyn, D. P., (Ed.) New York: Academic Press, 1967

Stubblefield, E.: The structure of mammalian chromosomes. Int. Rev. Cytol. 35, 1–60 (1973)

Sugawara, S., Hafez, E. S. E.: Developmental changes in dehydrogenase activities in rabbit eggs. Proc. Soc. Exp. Biol. 126, 849–853 (1967)

Sugawara, S., Umezu, M.: Studies on the metabolism of the mammalian ova. II. Oxygen con-sumption of the cleaved ova of the rat. Tohoku J. Agric. Res. 12, 17–28 (1961)

Surani, M. A. H.: Zona pellucida denudation, blastocyst proliferation and attachment in the rat. J. Embryol. Exp. Morphol. 33, 343–353 (1975)

Sydow, H.: Elektronenmikroskopische Untersuchung über zytosomale Lamellenkörper in den Ei-zellen des Ovars vom Igel (Erinaceus europaeus L.). Z. Zellforsch. 88, 387–407 (1968)

Szollosi, D.: Cortical granules: a general feature of mammalian eggs? J. Reprod. Fertil. 4, 223–224 (1962)

Szollosi, D.: Extrusion of nucleoli from pronuclei of the rat. J. Cell Biol. 25, 545–562 (1965a)

Szollosi, D.: Development of yolky substance in some rodents eggs. Anat. Rec. 151, 424 (1965b)

Szollosi, D.: Nucleolar transformation and ribosome development during embryogenesis of the rat. J. Cell Biol. 31, 1115 A (1966)

Szollosi, D.: Development of cortical granules and the cortical reaction in rat and hamster eggs. Anat. Rec. 159, 431–446 (1967)

Szollosi, D.: Nucleoli and ribonucleoprotein particles in the preimplantation conceptus of the rat and mouse. In: The Biology of the Blastocyst. Blandau, R. J., (Ed.) Chicago and London: The University of Chicago Press, 1971

Szollosi, D.: "Periodisch structurierte Körper" (PSK) in the perivitelline space of rat and mouse embryos. J. Ultrastruct. Res. 53, 222–226 (1975a)

Szollosi, D.: Mammalian eggs aging in the fallopian tubes. In: Aging Gametes. Blandau, R. J., (Ed.) Basel and New York: S. Karger, 1975b

Szollosi, D.: Hunter, R. H. F.: Ultrastructural aspects of fertilization in the domestic pig: sperm penetration and pronucleus formation. J. Anat. 116, 181–206 (1973)

Šťastná, J.: Changes in the submicroscopic structure of the rat blastocyst in the preimplantation period. Folia Morphol. (Prague) 20, 124–125 (1972)

Šťastná, J.: Origin and function of multivesicular bodies in the segmenting ovum of rat. Acta Fac. Med. Univ. Brun. 49, 87–98 (1974a)

Šťastná, J.: Evidence of lysosomal nature of cortical granules in rat ovum. Scripta Med. 47, 527–533 (1974b)

Šťastná, J.: Relationship of cortical granules to lysosomes in the rat ovum. Folia Morphol. (Prague) 22, 234–235 (1974c)

Šťastná, J.: Occurrence and location of acid phosphatase in rat ovum during cleavage. Scripta Med. 50, 21–34 (1977a)

Šťastná, J.: Occurrence and localization of succinic dehydrogenase in the rat ovum in the preimplantation period. Acta Fac. Med. Univ. Brun., in press (1977b)

Šťastná, J.: Personal communication (1977c)

Šťastná, J.: Golgi complex in rat ovum during cleavage (a morphological and cytochemical study). Scripta Med. 51, 31–38 (1978a)

Šťastná, J.: Submicroscopic localization of succinate dehydrogenase activity in the mitochondria of the segmenting rat ovum. Acta Univ. Carol. Med., in press (1978b)

Šťastná, J.: Fine structural localization of alkaline phosphatase in rat ovum during cleavage. Z. Mikrosk. Anat. Forsch., in press (1979)

Tasca, R. J., Hillman, N.: Effects of actinomycin D and cycloheximide on RNA and protein synthesis in cleavage stage mouse embryos. Nature (London) 225, 1022 (1970)

Thiéry, J.-P.: Mise en évidence des polysaccharides sur coupes fines en microscopie électronique. J. Microsc. 6, 987–1018 (1967)

Thomson, J. L., Biggers, J. D.: The effect of inhibitors of protein synthesis on the development of mouse embryos in vitro. Exp. Cell Res. 41, 411–427 (1966)

Thomson, J. L., Brinster, R. L.: Glycogen content of preimplantation mouse embryos. Anat. Rec. 155, 97–102 (1966)

Trávník, P.: Evidence of Non-Specific Esterase in the Segmenting Rat Ovum. (In Czech). Thesis, J. E. Purkyně Univ. Brno (1976)

Trávník, P.: Study of the incidence and localization of lipids in the rat oocyte and cleaving ovum. Folia Morphol. (Prague) 25, 15–20 (1977a)

Trávník, P.: Changes of submicroscopical localization of non-specific esterase during the cleavage of rat ovum. (In Russian). Acta Fac. Med. Univ. Brun., in press (1977b)

Trávník, P.: Ultrahistochemical detection of non-specific esterase during cleavage of rat ovum. Acta Univ. Carol. Med., in press 1978

Trujillo-Cenoz, O., Sotelo, J. R.: Relationship of the ovular surface with follicle cells and origin of the zona pellucida in the rabbit oocytes. J. Biophys. Biochem. Cytol. 5, 347–350 (1959)

Tucker, E. B., Schultz, G. A.: Polypeptide synthesis and gene expression in preimplantation rabbit embryos. J. Cell Biol. 70, 192a (1976)

Unger, B., Dickson, A. D.: Effect of cycloheximide and actinomycin D on the mouse blastocyst untergoing the giant cell transformation. J. Anat. 108, 519–525 (1971)

Vivarelli, E., Siracusa, G., Mangia, F.: A histochemical study of succinate dehydrogenase in mouse oocytes and early embryos. J. Reprod. Fertil. 47, 149–150 (1976)

Wales, R. G.: Accumulation of carboxylic acids from glucose by the preimplantation mouse embryo. Aust. J. Biol. Sci. 22, 701–707 (1969)

Wales, R. G., Biggers, J. D.: The permeability of two- and eight-cell mouse embryos to L-malic acid. J. Reprod. Fertil. 15, 103–111 (1968)

Wales, R. G., Brinster, R. L.: The uptake hexoses by preimplantation mouse embryos in vitro. J. Reprod. Fertil. 15, 415–422 (1968)

Wales, R. G., Whittingham, D. G.: Metabolism of specifically labelled pyruvate by mouse embryos during culture from the two-cell stage to the blastocyst. Aust. J. Biol. Sci. **23**, 877–887 (1970)

Wales, R. G., Whittingham, D. G.: The metabolism of specifically labelled lactate and pyruvate by two-cell mouse embryos. J. Reprod. Fertil. **33**, 207 (1973)

Wartenberg, H.: Elektronenmikroskopische und histochemische Studien über die Oogenese der Amphibieneizelle. Z. Zellforsch. **58**, 427–486 (1962)

Wartenberg, H., Stegner, H.-E.: Über die elektronenmikroskopische Feinstruktur des menschlichen Ovarialeies. Z. Zellforsch. **52**, 450–474 (1960)

Weakley, B. S.: Electron microscopy of the oocyte and granulosa cells in the developing ovarian follicles of the golden hamster (Mesocricetus auratus). J. Anat. **100**, 503–534 (1966)

Weakley, B. S.: Investigations into the structure and fixation properties of cytoplasmic lamellae in the hamster oocyte. Z. Zellforsch. **81**, 91–99 (1967)

Weakley, B. S.: Comparison of cytoplasmic lamellae and membranous elements in the oocytes of five mammalian species. Z. Zellforsch. **85**, 109–123 (1968)

Weakley, B. S.: Further observations on a cytoplasmic structure in the hamster oocyte. Z. Zellforsch. **146**, 517–523 (1973)

Weitlauf, H. M.: Hormonal regulation of protein synthesis in mouse blastocysts. Anat. Rec. **163**, 282–283 (1969a)

Weitlauf, H. M.: Temporal changes in protein synthesis by mouse blastocysts transferred to ovariectomized recipients. J. Exp. Zool. **171**, 481–486 (1969b)

Weitlauf, H. M.: Protein synthesis by blastocysts in the oviducts and uteri of hypofysectomized mice. J. Exp. Zool. **176**, 35–40 (1971)

Weitlauf, H. M., Greenwald, G. S.: A comparison of ^{35}S methionine incorporation by the blastocysts of normal and delayed implanting mice. J. Reprod. Fertil. **10**, 203–208 (1965)

Weitlauf, H. M., Greenwald, G. S.: A comparison of the in vivo incorporation of ^{35}S methionine by two-celled mouse eggs and blastocysts. Anat. Rec. **159**, 249–254 (1967)

Weitlauf, H. M., Greenwald, G. S.: Influence of estrogen and progesterone on the incorporation of ^{35}S methionine by blastocysts in ovariectomized mice. J. Exp. Zool. **169**, 463–470 (1968)

Whitten, W.K.: Culture of tubal mouse ova. Nature (London) **177**, 96 (1956)

Whitten, W. K.: Culture of tubal ova. Nature (London) **179**, 1081 (1957)

Whittingham, D. G.: The failure of lactate and phosphoenolpyruvate to support development of the mouse zygote in vitro. Biol. Reprod. **1**, 381–386 (1969)

Wischnitzer, S.: An electron microscope study of cytoplasmic organelle transformations in developing mouse oocytes. Wilhelm Roux' Archiv **166**, 150–172 (1970)

Wischnitzer, S.: The submicroscopic morphology of the interphase nucleus. Int. Rev. Cytol. **34**, 1–48 (1973)

Woodland, H. R., Graham, C. F.: RNA synthesis during early development of the mouse. Nature (London) **221**, 327–332 (1969)

Wu, J. T., Meyer, R. K.: Ultrastructural changes of rat blastocysts induced by estrogen during delayed implantation. Anat. Rec. **179**, 253–272 (1974)

Yamada, E. T. M., Motomura, A., Koga, K.: The fine structure of the oocyte in the mouse ovary studied with the electron microscope. Kurume Med. J. **4**, 148–171 (1957)

Zamboni, L.: Ultrastructure of mammalian oocytes and ova. Biol. Reprod., Suppl. **2**, 44–63 (1970)

Zamboni, L., Mastroianni, L., Jr.: Electron microscopic studies on rabbit ova. I. The follicular oocyte. J. Ultrastruct. Res. **14**, 95–117 (1966a)

Zamboni, L., Mastroianni, L., Jr.: Electron microscopic studies on rabbit ova. II. The penetrated tubal ovum J. Ultrastruct. Res. **14**, 118–132 (1966b)

Zamboni, L., Bell, J., Baca, M., Mishell, D. R., Jr.: A penetrated human ovum studied by electron microscopy. Nature (London) **210**, 1373–1375 (1966a)

Zamboni, L., Mishell, D. R., Jr., Bell, J. H., Baca, M.: Fine structure of the human ovum in the pronuclear stage. J. Cell Biol. **30**, 579–600 (1966b)

Zeilmaker, G. H., Hulsman, W. C., Weinsinck, F., Verhamme, C.: Oxygen-triggered mouse oocyte maturation in vitro and lactate utilization by mouse oocytes and zygotes. J. Reprod. Fertil. **29**, 151–152 (1972)

Zetterqvist, H.: The ultrastructural organization of the columnar absorbing cells of the mouse jejunum. Thesis, Karolinska Institutet, Stockholm (1956)

Subject Index